2022 Banff Pork Seminar

Advances in Pork Production Volume 33

Online
5-7 January 2021

Editors:

Michael K. Dyck
Ben Willing
Ashley Steeple

ISBN: 978-1-7138-5756-3

Printed from e-media with permission by:

Curran Associates, Inc.
57 Morehouse Lane
Red Hook, NY 12571

Some format issues inherent in the e-media version may also appear in this print version.

University of Alberta
Department of Agricultural, Food and Nutritional Science
4-10 Agriculture/Forestry Centre, University of Alberta
Edmonton, AB, Canada T6G 2P5

Phone: (780) 492-3651
Email: pork@ualberta.ca

www.banffpork.ca

Additional copies of this publication are available from:

Curran Associates, Inc.
57 Morehouse Lane
Red Hook, NY 12571 USA
Phone: 845-758-0400
Fax: 845-758-2633
Email: curran@proceedings.com
Web: www.proceedings.com

Advances in Pork Production

Volume 33, 2022

banff pork seminar

Advances in Pork Production

Volume 33

Proceedings of the 2022 Banff Pork Seminar

Co-operating Agencies

Alberta Pork

Alberta Agriculture and Forestry

University of Alberta

Published by:
University of Alberta
Department of Agricultural, Food and Nutritional Science

ISSN: 1489-1395

Editor–Papers
Michael K. Dyck
Editor–Abstracts
Ben Willing
Copyeditor
Ashley Steeple

Conference Coordinator
Ashley Steeple

For information please contact our office at:

Telephone: (780) 492-3651
Email: pork@ualberta.ca
Website: www.banffpork.ca

Mailing Address:
Banff Pork Seminar
c/o Department of Agricultural, Food and Nutritional Science
4-10 Agriculture/Forestry Centre, University of Alberta
Edmonton, AB, Canada T6G 2P5

Preface

50th Years of Knowledge and Sharing

I'm very pleased to welcome all participants to this year's hybrid Banff Pork Seminar. First and foremost, to all the people that are here today with us in Banff for the first main pig industry event in Canada with live attendees since March 2020. Also, thanks for those joining us on our virtual platform. We wish you could have been able to make it to Banff this year. Last year marked the 50th anniversary of the Banff Pork Seminar. We didn't have the chance to celebrate this milestone due to the Covid pandemic. This year, the advisory committee decided to add some special features to the program in order to celebrate this great achievement for the Canadian swine industry. You will a chance to review the 50th years of the Banff Pork Seminar throughout a video that was made in recognition for some of the key people that had a large influence on the success of this annual event.

This is why we selected 50th Years of Knowledge and Sharing for the for this year's theme. This pretty much summarizes in a few keywords what the Banff Pork Seminar is all about. Gaining knowledge and sharing it with all our industries. The program this year has been put together in a way that we can have more knowledge and idea sharing, trying also to get more connected with the people that work day to day in our swine operation. We also wanted to address some of the theme that are of high interest in our evolving industry. Our seminar wants to reach out to the entire spectrum of our industry, including farmers, students, agri-business professionals, and academia across the industry.

As we go into another year of Covid, we as an industry, keep facing many challenges. Although pork prices had been, for the most, pretty good in 2021 for our North American industry, we have seen a significant increase in feed cost due to factors such as grain and protein price, but also due to the large jump in amino acids, vitamins and other feed additive prices. The lingering effect of Covid on labor and container cost has putted a lot of pressure on our supply chain and prices for oversea product. We still expect some turmoil an uncertainty as we get into 2022. With the lower pig prices that we have seen over the last 2-3 months, it is even more important to look at your production and improve efficiency.

This year's Banff Pork Seminar would not happen without the dedication and support of many throughout our industry. The advisory group offers a heartfelt thank-you to our generous sponsors, without whom this conference would not be possible. We would be amiss if we didn't acknowledge the dedication and commitment of Ashley Steeple, the BPS coordinator and planner extraordinaire, and Ruurd Zijlstra and Michael Dyck, program directors. Finally, thanks to the Advisory Committee, for continuing to bring forward new ideas that have resulted in an incredible agenda again for this year seminar.

To all participants, I wish you a warm welcome and I hope you will appreciate our program for this year's 2022 Banff Pork Seminar.

Sincerely
Dan Bussieres
Chair, 2022 Banff Pork Seminar Advisory Committee.

2022 BPS Advisory Committee

Chair

Daniel Bussières, Groupe Cérès., Levis, QC

Program Directors

Ruurd Zijlstra, Professor, Swine and Carbohydrate Nutrition, Dept. of Agricultural, Food and Nutritional Science, University of Alberta, Edmonton, AB

Michael Dyck, Associate Professor, Reproductive Physiology/Biotechnology, Dept. of Agricultural, Food and Nutritional Science, University of Alberta, Edmonton, AB

University of Alberta

Ben Willing, Assistant Professor and Canada Research Chair in Microbiology of Nutrigenomics, Dept. of Agricultural, Food and Nutritional Science, University of Alberta, Edmonton, AB

Alberta Agriculture and Forestry

Jennifer Hannesson, Livestock Traceability Section, Edmonton, AB

Alberta Pork

Javier Bahamon, Quality Assurance Coordinator, Edmonton, AB

Industry Representatives

Eric Aubin, DNA Genetics, St-Lazare, Que.

Javier Bahamon, Alberta Pork, Edmonton, Alta.

Jason Brandt, Steve's Livestock Transport, Red Deer, Alta.

Melissa Da Costa, Baseline Pork, Harley, Ont.

Jen Demare, Swine Heath Professionals, Steinbach, Man.

Cliff Ehr, Olymel, Humboldt, Sask.

Heather Ferguson, Cargill Animal Nutrition, Puslinch, Ont.

Michel Larivière, PIC, St. Malo, Man.

Frank Novak, Sunhaven Farms, Sherwood Park, Alta.

Rick Prejet, Porcherie Lac du Onze, Notre Dame de Lourdes, Man.

Cordell Young, Precision Veterinary Services, Lethbridge, Alta.

Conference Coordinator

Ashley Steeple, University of Alberta, Edmonton, AB

We extend sincere thanks to the following companies and groups who have provided financial assistance to the 2022 Banff Pork Seminar and to the publication of these proceedings.

Sustaining Sponsors

Alberta Pork
Fast Genetics
Maximus Systems
PIC
Swine Innovation Porc
Topigs Norsvin Canada

Premier Plus Sponsors

Chr. Hansen
Dept. of Agricultural, Food and Nutritional Science, University of Alberta
Hypor
Maple Leaf
New Standard West Equipment

Premier Sponsors

Bio Agri Mix & Phibro Animal Health
Canadian Pork Councils
Cargill
DSM Nutritional Products
HyLife
Jefo
Nutrition Partners Inc. / Farmhouse Consultants Inc.
Penner Farm Services
South West Vets
Steve's Livestock Transport

... *Sponsors of the 2022 Banff Pork Seminar (continued)*

Bronze Sponsors

Agribution Canada Ltd.
Alltech
Animal Science Products Inc.
APC.
Automated Production
Barkman Concrete Ltd.
Boehringer Ingelheim Animal Health Canada Inc.
Champion Alstoe Animal Health
DNA Genetics
EastGen
Farm Credit Canada
FeedFlo
Gestal Jyga Technologies
Grand Mfg Co. Ltd.
Grand Valley Fortifiers
h@ms Marketing Services
Hipra Animal Health
Innovative Heating Technologies
Kemin
Minitube Canada
Modern Veterinary Therapeutics
Neogen
Ontario Pork
Polar Pork Farms
Prairie Hog Country
S.E.C Repro inc.
Walbern Agri-Systems
Western Union Business Solutions (Custom House ULC)
Zoetis

Index of Advertisers

Program of the 2022 Banff Pork Seminar
January 11-13, 2022

Plenary Sessions

Wednesday

- **Markets & Trade**

  Brett Stuart, *Global AgriTrends*

- **Positioning Canadian Pork in a Post Covid-19 Environment**

  Trevor Sears, *Canada Pork*

Thursday

- **Agriculture & the Environment**
  Patrick Moore, *Greenspirit*

- **Successful Strategies Used to Prepare for African Swine Fever in Canada**
  Christa Arsenault, *National Farmed Animal Health & Welfare*

- **Views from an Agricultural Immigrant**
  Vincent ter Beek, *Pig Progress*

Breakout Sessions

1. **Water Quality**

- **Getting the Most From Your Water Program**

  Susan Watkins, *University if Arkansas*

- **Do You Know About Your In-Barn Water Quality & How to Improve It**

  Martine Giguere, *Agro-Bio Controle Inc.*

2. **Feeding**

- ❖ **Precision Nutriton**

  Brent Frederick, *Provimi*

- ❖ **Reducing Feed Costs in Grow-Finish: Alternative Feedstuffs & Production Strategies**

  Eduardo Beltranena, *University of Alberta*

<u>Breakout Sessions</u>

3. **Human Resources**

- ❖ **Love the Ones You're With… Staff Training & Development**

  Dennis Robles, *Swine Health Professionals*

- ❖ **When Stress is More Than a Season**

  Lesley Kelly, *High Heels & Canola Fields*

4. **Innovators**

- ❖ **Disease-causing Brachyspira hampsonii efficiently harvest iron, likely outcompeting beneficial bacteria**

  Hannah Burlet, *University of Saskatchewan*

  The effect of limiting dietary non-essential amino acids on lysine requirement for protein deposition in growing pigs

  Carley Camire, *University of Saskatchewan*

  Immunological evaluation of autogenous vaccine used in sows to protect piglets against Streptococcus suis infection

  Alison Jeffery, *University of Montreal*

FX Aherne Award for Innovative Pork Production

- ❖ **Back spacer device**

  Jewel White, *Maple Leaf Agri-Farms*

Table of Contents

Plenary Sessions

Breakout Sessions

1. Water Quality

5. Management & Production

6. Sow Productivity

7. Welfare

8. Health

WE'RE WITH
YOU EVERY DAY
At Fast Genetics we combine the latest technology and innovation with generations of knowledge to create superior products for producers, and in the process, we redefine the everyday - every single day. The future of pork production starts with you, and we are committed to walking step by step beside you to build it together.
FIND OUT HOW WE ARE COMMITTED TO PRODUCERS AT FASTGENETICS.COM
Fast Genetics™

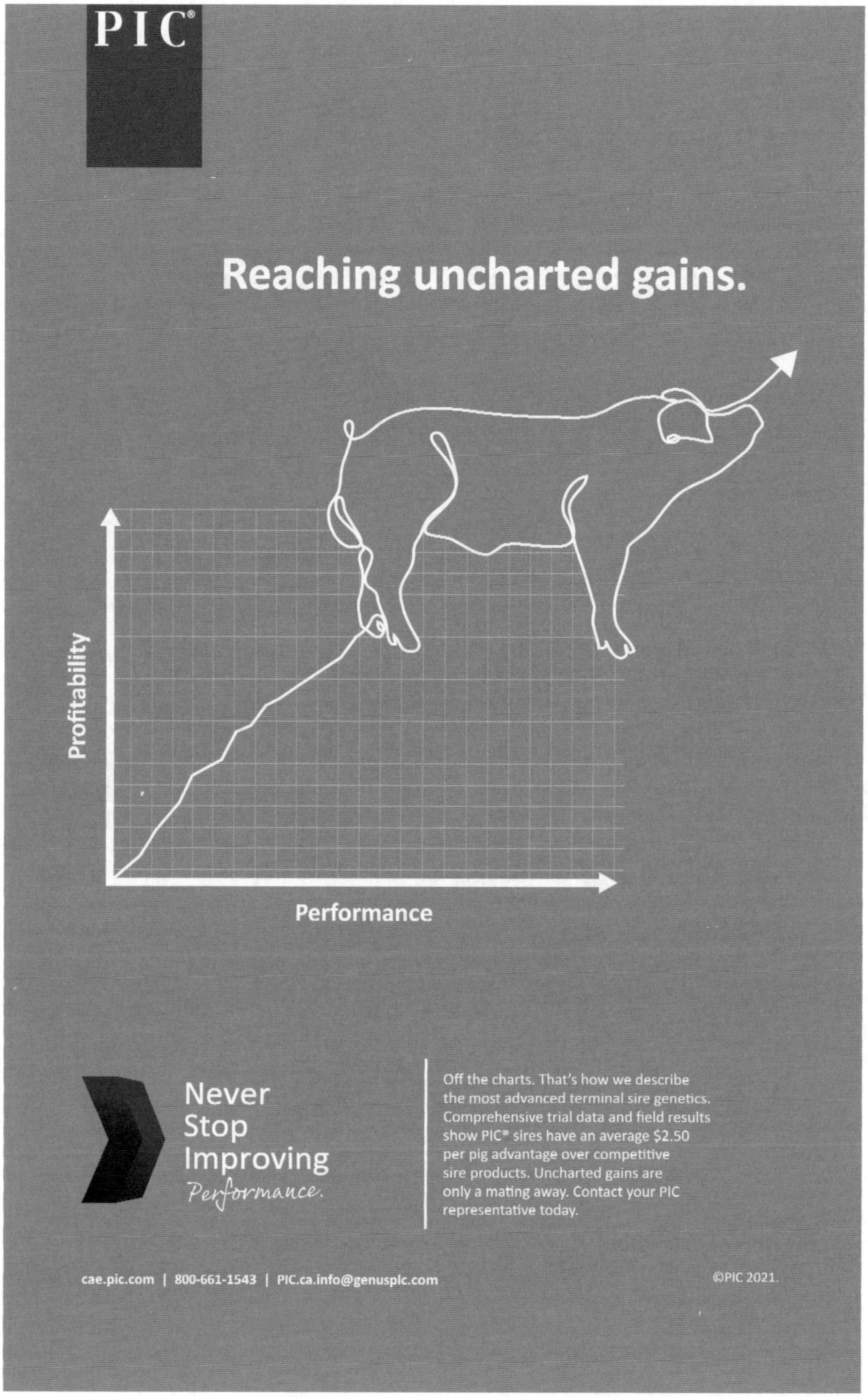

PIC®
Reaching uncharted gains.
Profitability
Performance
Never
Stop
Improving
Performance.
Off the charts. That's how we describe the most advanced terminal sire genetics. Comprehensive trial data and field results show PIC® sires have an average $2.50 per pig advantage over competitive sire products. Uncharted gains are only a mating away. Contact your PIC representative today.
cae.pic.com | 800-661-1543 | PIC.ca.info@genusplc.com
©PIC 2021.

TN70
Parent sow
Make your
business

Global Pork Market Driven by COVID-19, ASF, and Rising Input Costs

Brett W. Stuart

bstuart@globalagritrends.com, 1022 W. 3200 S. Preston, Idaho 83263

1. Macroeconomic Factors Affecting Global Pork Markets

Economic growth, per capita income, money supply, and currency rates have always underpinned global pork markets. But the past two years have brought an unprecedented influence. Social and government responses to COVID-19 snarled logistics and affected both consumers' supplies and demand. These responses also spurred global commodity inflation higher, with shifts in both supply and demand.

1.1 COVID-19 and the Economy

COVID-19 was a classic "Black Swan" event. Unprecedented, unexpected, random, and impossible to forecast. The world watched as deaths in Wuhan and surrounding Hubei province, China soared above 68,000 cases[i]. As cases spread globally in January 2020, it became very apparent that the virus would sweep the earth.

What was amazing then and still is today, is that the virus in Wuhan initially showed survivability rates of 99% or higher. On a global basis today the rate is 98%[ii], but considering that cases are under-reported (only official cases are reported), the likelihood is that the survivability rate of those infected with COVID is above 99%.

Nevertheless, government intervention soared as consumer fears rose. Within 2 months of the pandemic, US$10 trillion in global economic stimulus was announced.[iii]

The U.S. money supply swelled. By March 2021, the U.S. "M2" money supply was up +27% from the prior year[iv]. Federal Reserve Bank Chairman Jerome Powell testified to congress that this increase was "a poor indicator of inflation" and that we must "unlearn" that correlation. Mr. Powell rejected the notion that inflation concerns outweighed the COVID-19 concerns. Within a month the Federal Reserve stopped reporting the "M2" dataset of U.S. money supply. Another indicator reported by the Federal Reserve, Deposits in All Commercial Banks, jumped from US$13.3 trillion in January 2020 to US$17.5 trillion in October 2021[v].

1.2 Inflation and the Commodity Super Cycle

The combined effect of this heavy stimulus increased consumer spending, creating strong demand responses for commodities globally. At the same time, manufacturing and processing slowed as sick workers reduced productivity. Also, China became a strong buyer of pork and corn due to their African Swine Fever (ASF) shortages and ban on swill feeding. The combined effect was the beginning of a commodity super cycle.

Global inflation soared through the pandemic but was most pronounced in 2021. By May 2021 commodity inflation had surged higher on drought concerns, global shipping backlogs, and soaring U.S. consumer demand. Real estate prices advanced higher and stock markets continued besting record highs.

A commodity super cycle is created when broad factors, both internal and external, combine to create sustained higher price levels. Global commodities prices had been subdued from 2015 to 2019. Chinese crop demand, drought fears in both South and North America, were the internal factors along with China's ASF-driven hog culling. The tinder was set and COVID-19 stimulus, demand growth, and supply chain kinks, were the spark.

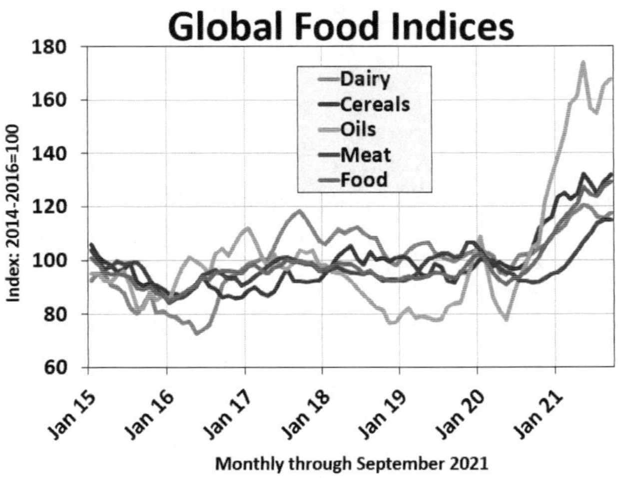

Figure 1. Global Commodity Indices
Source: UN / Food and Agriculture Organization

2. Fundamental Supply/Demand Factors Affecting Global Pork Markets

2.1 U.S. Swine Supply and Demand

U.S. hog producers saw decent profits for the first time in two years beginning in October 2020. Those profits dipped over the winter and then surged higher into

2021 with profits nearly triple the 5-year average from April through July. We estimated peak producer profits near $54 per head in June 2021.

Supplies were tightened due to some more severe strains of a Porcine Reproductive and Respiratory Syndrome (PRRS) virus. That combined with strong consumer demand and good exports in early 2021, drove producer profits. Typically, this kind of profit drives production growth. That was exactly what happed after the 2014 markets surged on tight supplies from Porcine Epidemic Diarrhea (PED) outbreaks. We saw production surge higher, along with 4 new producer-owned (or affiliated) processing plants. But that was not the case in 2021.

As lumber prices moved to new record highs, labor shortages emerged. Concerns over slowing China pork demand and uneasiness over disease outbreaks, expensive materials and labor led producers to hold back on expansion. The September 2021 USDA Hogs and Pigs report showed total inventories down -4%, breeding hogs down -2%, and the pig crop down -4% from the prior year.

2.2 China's Swine Situation / ASF Update

China's ASF saga started with discoveries in August 2018, leading to what we believe was a swine cull of 65% in 2019 and 2020. The resulting pork shortages drove hyper-inflation through China's hog and pork sector. Chinese consumers were paying 140% more for pork in March 2020 compared to the prior year.

Figure 2. Retail/Wholesale China pork prices
Source: China Ministry of Agriculture and Rural Affairs

Those prices ebbed and flowed but remained incredibly high for nearly two years. As a result, Chinese hog farmers saw hog production margins above US$200/head for 20 months, a level beyond imagination, that would ultimately cause problems down the road.

By January 2021, ASF outbreaks appeared to be increasing. More unofficial reports suggested widening outbreaks particularly in north and central China. From January 2021 to mid-June, hog prices across China had fallen -60% and pushed margins back below breakeven. What started as disease liquidation, turned into financial liquidation. Prices languished at loss-levels through October. There were reports of feed companies not being paid and even villagers releasing hogs that they could no longer afford to feed.

China's liquidation in 2021 was driven by both disease culling and financial liquidation. We expect sharp price inflation for Chinese hogs and pork into 2022.

This will likely lead to increased Chinese pork import demand in the 2nd half of 2022.

2.3 European Swine Situation

European hog farmers are feeling the market pressure from China's slowdown in demand. Prices have shifted lower moving most producers into losses across Europe. The China slowdown combined with feed price inflation will likely drive liquidation of hogs in Germany and Denmark. Spain remains the #1 pork producer and exporter in Europe but has also felt the sting of sharply lower exports. As a result European hog inventories will likely contract in 2022.

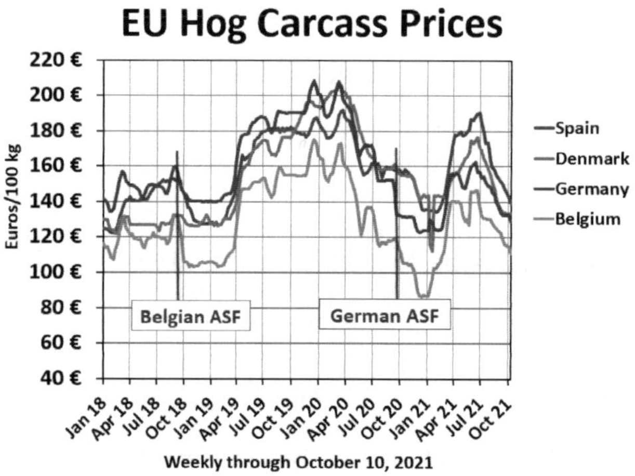

Figure 3. .EU Hog Prices
Source: BPEX

African Swine Fever (ASF) remains a serious hindrance for European pork producers. While Belgium was able to stamp out their wild boar outbreak, export bans remain. They recently were approved to export to Japan after more than one year without any new cases. But Germany faces a much more difficult future.

They have found more than 1,000 infected wild boars and ASF was found on two domestic German hog farms. Thus, they are OUT of the global markets. They can export within the EU but cannot export externally. And with continued findings, they will likely be out of the key export markets for years.

3. Other Variables to Watch in 2022

3.1 ASF on the Move

ASF remains a serious threat to pork producers all over the world. In August, positive ASF cases were found in the Dominican Republic and then later in September they were reported in Haiti. These are the first cases in the Western hemisphere in roughly 40 years. It remains a mystery how the virus crossed an ocean. Dominican hog farmers reported that Haitians came over in late 2020 to buy swine, reporting that many of theirs had died from a virus. ASF may have been in Haiti much longer than initially feared.

The risk is obvious. There is a constant flow of goods and people among the Caribbean basin and the U.S. In fact, if a positive ASF case were found in Puerto Rico, just 80 miles from the Dominican Republic, any global pork importer would be justified in banning U.S. pork exports. The USDA has pre-emptively moved to designate Puerto Rico a “protective region”. That would remove the justification to ban mainland exports were case to be found there. But that process takes time. ASF cases continue to be reported in Russia, Africa, Asia, and Eastern Europe. This virus is pernicious and difficult to eradicate, particularly when it reaches the wild boar populations.

3.2 Rising Input Costs

Higher feed grain and soy prices began shifting higher in late 2020 based on drought concerns in both North and South America, as well as strong Chinese buying. Even after a correction from the May 2021 peak but remain elevated.

Drought concerns remain as a strong La Nina pattern continues in the Pacific Ocean. But China will likely need less feed in 2022 based on the heavy swine herd liquidation of 2021. And Chinese officials continue to suggest one of the largest crops produced in 2021.

However, inflation concerns persist with sharply higher and volatile fertilizer prices.

4. Conclusion

The external events of recent years, including COVID-19 and ASF have reshaped not only global pork, but global commodity markets. Government responses have topped up money supplies creating an inflationary hike unlikely to be undone. Volatility has increased on tight supplies of materials and labor. With ASF still on the move, and potentially threatening the Western Hemisphere, 2022 will be marked by increased market volatility. But reduced swine inventories in China and Europe bode well for global pork producers.

References

i. Johns Hopkins University Center for Systems Science and Engineering (JHU CSSE).

ii. Ibid.

iii. Ziyad Cassim, Borko Handjiski, Jörg Schubert, and Yassir Zouaou, June 2020, McKinsey and Company, The $10 trillion rescue: How governments can deliver impact

iv. https://fred.stlousfed.org, U.S. Federal Reserve Bank

Ibid.

Positioning Canadian Pork in a Post COVID-19 Environment

Trevor Sears

Canada Pork, 220 Laurier Avenue West, Suite 900, Ottawa, Ontario, K1P 5Z9
Email: sears@canadapork.com

Introduction

Canada is a major global player in premium quality pork production. Despite some of the challenges the industry has faced during the pandemic, 2020 was a record year for Canadian pork with respect to production and export volumes. Canadian pork production in 2020 reached 2.3 million tonnes, and amidst supply side challenges in many countries around the world, pork from Canada has been in demand more than ever, representing more than 1.4 million tonnes being exported to over 90 countries.

However, the strength in Canadian exports combined with pandemic restrictions placed on Canadian foodservice establishments, and a situation that has found more people working from home has undoubtedly shifted buying behaviour amongst Canadian consumers. As a result, the closest measure of pork consumption in Canada we have (domestic pork disappearance per capita) declined significantly during the pandemic (down 13.9% from 2019).

Post pandemic, the Canadian pork industry will continue to face many challenges. Market access issues, changing political agenda and/or agriculture policy, food safety, herd health issues or biosecurity, can cause trade disruption. The Canadian pork industry exported over 70% of its production last year and the vast majority of that volume goes to Canada's top ten export markets. Significant disruption in trade to any of these markets could trigger a huge economic impact on the Canadian pork industry. These circumstances represent a serious threat to Canada's largest meat industry.

Positioning Canadian Pork in a post COVID-19 environment will require a robust and multi-faceted strategy. Among this includes promoting the premium quality value proposition of the Canadian pork industry at home and around the globe, which is the combination of Canada's systems for on-farm food safety and quality assurance, coupled with world-class federally inspected processing plants across Canada maintaining the highest standards for freshness, quality and safety. These systems play an integral part in securing and maintaining market access for Canadian pork products.

Background

The Canadian pork industry has enjoyed considerable growth over the last three to four decades, due in large part to a focus on developing export opportunities for Canadian pork products by improvements in swine genetics, by farmers who take pride in food production, and a solutions based processing culture combined with outstanding meat quality and workmanship to meet the needs of the most demanding markets.

Thirty years ago (1990), the Canadian pork industry exported approximately 26% of its production, representing just over 266,000 tonnes, mostly to the United States (75%).

Today, Canada is the world's third largest exporter of pork behind the EU and USA. Canada exports 71% of its pork production, higher than any other country, including Chile (50%), Brazil (29%), the United States of America (25%), or Mexico (24%).

The Canadian industry has cultivated a range of diverse market destination opportunities for pork products. In 2020 the top five largest markets for Canadian pork exports were China (610,000 tonnes), the United States (311,000 tonnes), Japan (255,000 tonnes), Mexico (107,000 tonnes), and the Philippines (48,000 tonnes). South Korea and Vietnam are also important market destinations for Canadian pork by volume.

Canada Pork Programs & Services for the Canadian Pork Industry

Canada Pork is the domestic and global market development and promotion organization for the Canadian pork industry.

Our members are united in the belief that market development, expanded market access opportunities and consumer confidence are essential for the growth and prosperity of the Canadian pork industry. Industry investment leveraged against innovative programs and services has been the recipe for success used by Canada Pork in the export and domestic marketplace for almost 30 years.

On behalf of members, Canada Pork works to improve market access for all Canadian pork products in part by making the resolution of market access issues a priority through the establishment of partnerships between industry and government, while promoting and supporting multilateral and bilateral trade agreements for the pork industry.

In addition, Canada Pork programs and services enhance the competitive position of Canadian produced pork by building product knowledge in all markets of importance through education, information, and value-added programs and services.

Branding Canada – Verified Canadian Pork Global Branding Strategy

Officially launched in 2015, the Verified Canadian Pork brand represents the very best combination of attributes that Canada can put in the box based on on-farm food safety, traceability and responsible animal care, coupled with Canada's world-class in-plant food safety and quality assurance systems. The VCP brand has been identified by Canada Pork as one of the key strategic initiatives in differentiating Canadian Pork on the world marketplace with successful launches in Canada, Japan, China, South Korea, Mexico, and elsewhere.

As a result, the on-farm food safety and quality assurance systems that Canadian farmers have followed for years combined with reliable, safe processing systems based on HACCP and a world class food inspection system are promoted directly to consumers in Canada and around the world.

Generic Promotion & Research for the Domestic Market

The creation of the new Pork Promotion and Research Agency is good news for consumers and all pork industry supply chain participants in Canada. Once fully implemented, the Agency will provide enhanced support for the domestic market with respect to generic pork promotion and research activities, which is greatly needed to level the playing field in Canada. In order to respond to a changing consumer landscape, it's crucial that expanded opportunities be made available as a part of an overall strategy to educate consumers more broadly about cooking and preparation techniques for pork, as well as the nutritional benefits of including pork on menus.

Other Global Factors

1. African Swine Fever
    a. Canada Pork will play a critical role in providing accurate and timely information to end-user customers in Canada and around the world in the event that African Swine Fever should be detected in Canada. This is crucial to maintaining and supporting consumer confidence of Canadian produced pork products. In addition, the Canadian pork sector in collaboration with industry and

government partners is working steadfastly to secure zoning arrangements with trading partners to ensure continuity of commercial pork trade.

2. Global Market Access Opportunities & Challenges
    a. Last year, China was Canada's largest pork customer and they are importing record high protein volumes from almost all exporting countries – and growing. In spite of this, Canada finds itself with a large portion of its production suspended on the pork side. This reduced access could cost the red meat industry well over $ 1 billion per year of exports.
3. Global supply chain and labour issues
    a. Labour and supply chain issues are not a Canadian issue alone rather one that is being experienced globally.

Conclusion

Competition is fierce as countries look to expand their share of the export and domestic marketplace. In a post-pandemic environment continued investment is crucial in the development and promotion of a "Branding Canada" approach across all markets and consumer segments to earn consumer trust and build brand equity in Canadian pork products. Effective market development programs and services coupled with robust, accurate and timely market information is required to maximise marketing and promotion opportunities for Canadian companies.

Fake Invisible Catastrophes and Threats of Doom

Patrick Moore

It dawned on me one day that most of the scare stories in the media today are based on things that are either invisible, like CO2 and radiation, or very remote, like polar bears and coral reefs. Thus, the average person cannot observe and verify the truth of these claims for themselves. They must rely on activists, the media, politicians, and scientists - all of whom have a huge financial and/or political interest in the subject - to tell them the truth. This is my effort, after 50 years as a scientist and environmental activist, to expose the misinformation and outright lies used to scare us and our children about the future of the Earth. Direct observation is the very basis of science. Without verified observation it is not possible to know the truth.

Citizens are being misinformed by many environmental doomsday prophesies, ones they cannot verify for themselves. We are told that nuclear energy is very dangerous when the numbers prove it is one of the safest technologies. We are told polar bears will go extinct soon when their population has been growing steadily for nearly 50 years. We are told that there is something harmful in genetically modified food crops when it is invisible, has no name and no chemical formula. We are told severe forest fires are caused by climate change when they are actually caused by poor management of fuel load (dead wood) in the forest.

We are told that all the coral reefs will die by 2100 when in fact the most diverse coral reefs are found in the warmest oceans in the world. And of course, we are told that invisible CO2 from using fossil fuels, accounting for more than 80 percent of our energy supply, will make the Earth too hot for life. All of these scare stories, and many more, are simply not true. There is no substitute for the truth.

Dr. Patrick Moore was one of the co-founders of Greenpeace and sailed on the first Greenpeace campaign against US H-bomb tests in Alaska. Upon receiving his PhD in ecology, he spent 15 years in the top committee of Greenpeace and led many of its environmental campaigns. Greenpeace began as a group of volunteers with noble intentions. Over the years it became very successful with campaigns to save the whales, stop the mass slaughter of baby seals, prevent toxic dumping into the air, water and earth, and many more. Greenpeace found itself in the early 1980s with more than $100 million coming in annually and close to 1,000 people on the payroll. It had become a business and fundraising moved to the top of the priority list. New campaigns were more about using sensationalism, misinformation and fear to attract donations.

Dr. Moore said good-bye in 1986 as Greenpeace was turning into a racket peddling junk science. Since then he has strived to be a sensible environmentalist, basing his beliefs on sound science and logical thinking. His work is the culmination of 50 years of learning during Dr. Moore's multi-faceted quest for the truth about environmental issues (an historical account of Dr. Moore's 15 years with Greenpeace and his analysis of environmental subjects are in his previous book,

"Confessions of a Greenpeace Dropout - The Making of a Sensible Environmentalist".

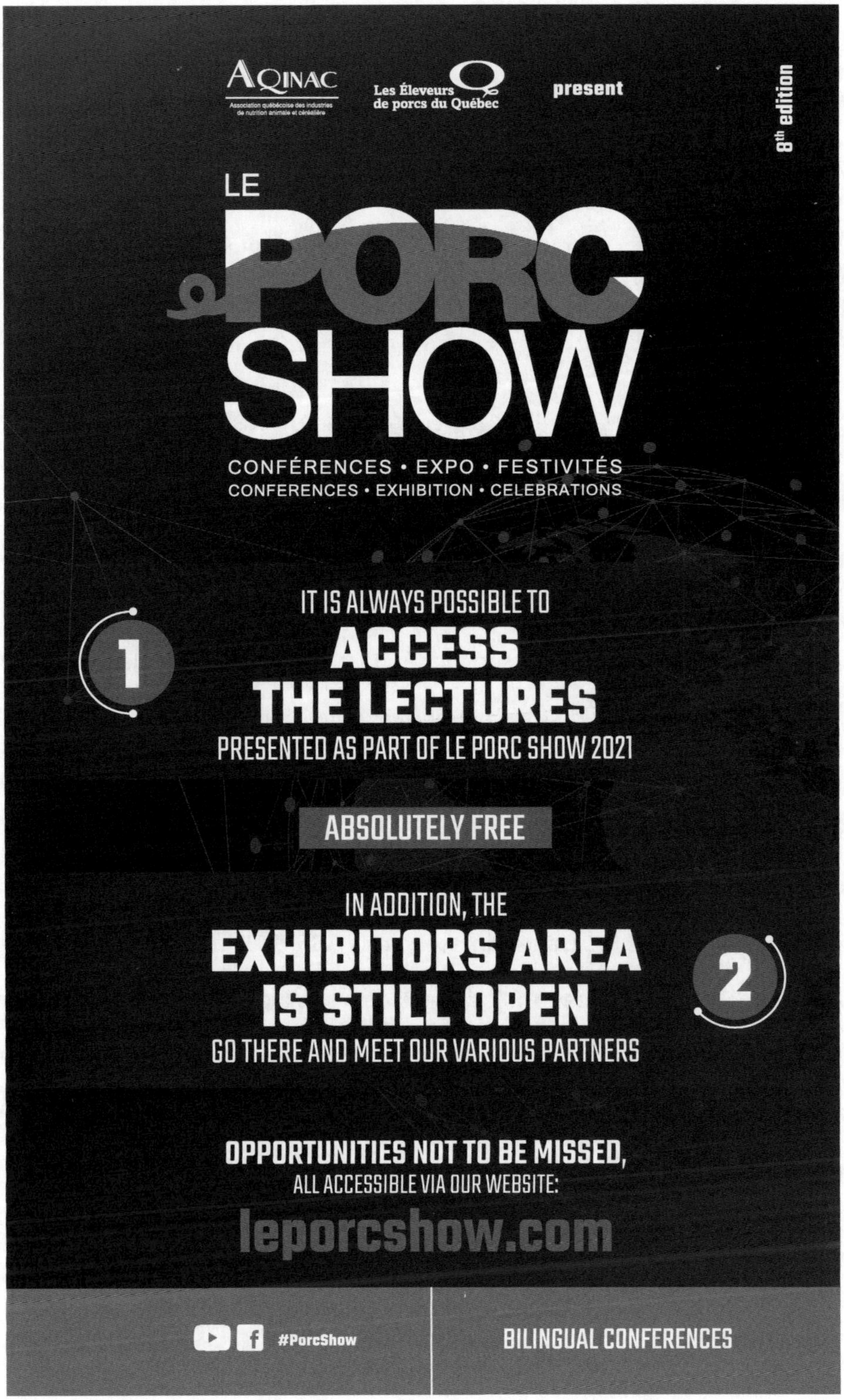
AQINAC
Association québécoise des industries de nutrition animale et céréalière
Les Éleveurs de porcs du Québec
present
8th edition
LE PORC SHOW
CONFÉRENCES • EXPO • FESTIVITÉS
CONFERENCES • EXHIBITION • CELEBRATIONS
IT IS ALWAYS POSSIBLE TO
1
ACCESS THE LECTURES
PRESENTED AS PART OF LE PORC SHOW 2021
ABSOLUTELY FREE
IN ADDITION, THE
EXHIBITORS AREA IS STILL OPEN
2
GO THERE AND MEET OUR VARIOUS PARTNERS
OPPORTUNITIES NOT TO BE MISSED,
ALL ACCESSIBLE VIA OUR WEBSITE:
leporcshow.com
#PorcShow
BILINGUAL CONFERENCES

Successful Strategies Used to Prepare for African Swine Fever (ASF) in Canada

Christa Arsenault[1], Morgan Radford[2] and Michelle Thompson[3]

[1]DVM, National ASF Coordinator- National Farmed Animal Health and Welfare Council; [2]BSC (AGR), ASF Project Lead- Animal Health Emergency Management; [3]DVM, MPH, ASF Project Lead- National Farmed Animal Health and Welfare Council

1. Background Information on African Swine Fever (ASF)

African Swine Fever (ASF) is a virus that only impacts pigs. This virus was first discovered in Africa in 1907 and was relatively stable for a century. In 2018 this changed with an outbreak of ASF being detected in China that quickly spread to other countries in Asia and Europe. In August 2021, ASF was detected in the Dominican Republic, the first country in the Americas confirmed to be infected since the virus began to spread globally in 2018. In September 2021, Haiti was also confirmed to be infected with ASF virus. ASF has never been detected in Canada, but it poses a significant risk to the health of the Canadian swine herd, the pork industry, and the Canadian economy. ASF does not infect humans and Canadian pork is safe to eat. ASF spreads between pigs through both direct and indirect contact with other infected pigs or pork products, as well as contaminated farm equipment, feed, and clothing. In Canada, ASF is a reportable disease under the *Health of Animals Act*. All suspected cases must be reported immediately to the Canadian Food Inspection Agency (CFIA).

2. Introduction

The ASF Executive Management Board (ASF EMB) was established in January 2019 to enable planning and preparedness efforts in advance of ASF being detected in Canada, as well as to coordinate efforts. The ASF EMB includes members from the following organizations:

- The Canadian Food Inspection Agency (CFIA)
- The Canadian Pork Council (CPC)
- Agriculture and Agri-Food Canada (AAFC)
- Canadian Meat Council (CMC)
- National Farmed Animal Health and Welfare Council (NFAHWC)
- Canadian Border Services Agency (CBSA)
- Council of Chief Veterinary Officers (CCVO)
- Provincial agricultural ministries

The ASF EMB also has ex-officio representation from:

- Environment and Climate Change Canada (ECCC)
- Canadian Wildlife Health Cooperative (CWHC)
- Provincial and Territorial Wildlife Directors

All ASF EMB members have been working together to advance the Pan-Canadian ASF Action Plan. The Plan includes priority activities organized under four framework pillars:

- Pillar 1: Enhanced biosecurity and prevention (lead by the CPC)
- Pillar 2: Preparedness and planning (lead by the CFIA)
- Pillar 3: Ensure business continuity (lead by the CMC)

- Pillar 4: Coordinated risk communications (lead by the Ontario Ministry of Agriculture, Food and Rural Affairs (OMAFRA))

In addition to the ongoing work by the ASF EMB, many provinces, provincial swine industry associations, provincial pork organizations, processors and individual swine operations have developed internal structures and groups to work on ASF planning and preparedness initiatives.

One of the biggest challenges faced by countries with ASF infections to date is how to control the spread of the virus once it is detected in the wild pig population. Wild pigs not only increase disease spread, but they are an invasive species that destroys environmental habitats, agriculture crops and pose a safety threat to humans and other native species. Since wild pigs exist in Canada, the ASF EMB is working closely with the Canadian Wildlife Directors Committee (CWDC), and the CWHC to ensure preparedness for wild pig sampling and surveillance, best management practices and wild pig-specific communications. Under the direction of the ASF EMB, the Canadian Invasive Wild Pig working group launched in April 2021 with the goal to develop a Canadian Wild Pig Strategy that will fully support the eradication of wild pigs as an invasive species. This national strategy will attempt to incorporate already developed provincial/regional wild pig control and elimination efforts within its framework.

3. Successful Strategies

The ASF EMB is setting the stage for collaborative planning between federal and provincial/territorial governments and swine industry groups in advance of a foreign animal disease being detected in Canada. This is the first time that this has been attempted in Canada in advance of response efforts. Since this is uncharted territory, there have been many lessons learned along the way. There have been many successful strategies implemented that have resulted in the development of actioning what is required to prepare Canada for ASF. Below are the top 4 successful strategies that have enabled an all-Canadian team-based approach and assisted those involved in working together towards the same goals. The intent of sharing these strategies is so that others can learn from our experiences. Since ASF planning and preparedness is an ongoing effort, examples of accomplishments and products produced will be included in the oral presentation that will be given at the Banff Pork Seminar in January 2022.

3.1 The Importance of Coordination

The ASF EMB identified a need for improved coordination between different pillars and amongst working groups to prevent duplication of tasks and ensure coordination in a forward direction. Dr. Christa Arsenault was hired into the national ASF coordinator position on April 1, 2021. Right away, it was identified that in order to provide coordination to all involved in this file, more resources would need to be hired. As of June 2021, the ASF coordination team consists of 3 people (1 full-time and 2 part-time equivalents).

Early on, the ASF coordination team reached out to provincial Chief Veterinary Officers (CVOs) and to the provincial pork producer associations' general managers to gain a better understanding of the following:

- The status of ASF planning at provincial levels; and
- Any gaps or challenges encountered to date that they felt needed to be addressed

From this reach out 2 main gaps were identified:

- A need for more communications from the ASF EMB; and
- A request for national coordination to plan for a swine welfare cull. Respondents wanted more information and national leadership from the perspective of timeliness, prioritization, financial assistance programs and clarity of roles and responsibilities ("who would do what?"). These comments have been actioned and the decision was made to modify an existing working group known as the Hog Supply Working Group, to address this gap as much as possible.

3.2 Communication, Communication, Communication

The importance of communications can never be overlooked. Both external/public facing communications and internal communications are vitally important. A key priority of the ASF EMB is to provide common and consistent messaging to all target audiences.

Pillar 4 of the ASF EMB is focussed on coordinated risk communication and leads external/public facing communication efforts. This team has successfully developed the following:

- Traveller awareness campaigns and travel industry outreach
- Biosecurity awareness campaign
- Detector dog promotions
- International outreach and promotion of zoning agreements
- Emergency preparedness communications and contact lists
- Creation of a public facing ASF newsletter.

Internal communications focussed on what was happening at the ASF EMB was left up to all represented organizations at the ASF EMB. It became vitally clear that routine communications to share ASF EMB key messages and accomplishments with provincial governments (especially those not involved directly at the ASF EMB level), provincial pork organization staff, swine industry associations and processors involved in ASF planning and preparedness initiatives was needed. Tools were developed for these routine communications that used a consistent ASF EMB brand with similar colour schemes, fonts and a logo that would resonate consistent messaging with the target audiences. A monthly meeting summary that captured relevant information from ASF EMB meetings was developed along with a quarterly ASF EMB pillar update to capture and communicate the progress of working groups. Both communications were posted on the NFAHWC website portal where they were made available to an approved recipient list.

The CPC also invited the NFAHWC ASF coordination team to provide monthly ASF planning and preparedness updates to the general managers and animal health leads of the provincial pork organizations. Repetitive messaging was deemed to be acceptable as sometimes the same message could be heard multiple times on multiple platforms.

Websites became increasingly important to act as the "go to place" to find ASF planning and preparedness information. Consistent messaging was emphasized and routine review and updating of websites was resourced and scheduled. The NFAHWC website was updated to include a section on ASF with a resource page linking to other ASF web resources. The NFAHWC content targets members of the ASF EMB, provincial government staff, provincial pork boards, swine industry staff and processors working on ASF planning and preparedness initiatives.

CPC, CMC, and Canada Pork collaborated to launch another new website with interactive videos on biosecurity and the importance of ASF prevention. This website targets Canadian swine producers, provincial pork organizations, pork consumers, food retail and restaurants.

Building strong internal and external communications by working collaboratively is a building block needed to be successful with communications involved in any animal health emergency response. This is a strategy that we continually strive to improve on.

3.3 Knowledge of Who is Doing What - The Importance of Detailed Org Charts

The ASF coordination team commonly receives questions from swine industry members and stakeholders asking: "What is the focus of each working group under the ASF EMB?" "Why are there so many working groups?" In response to these questions, the ASF Coordination team has been working on formatting and revising the ASF EMB organization chart. This document demonstrates the connection between each working group and the Pan-Canadian Action Plan's 4 pillars. It also includes all sub-working groups, documents the focus of each working group and the Chair's contact information. This document provided the knowledge needed to others that nothing was being overlooked and that repetition was being avoided as much as possible.

The ASF coordination team has also created a Wild Pig Org Chart that details all of the different working groups focussed on wild pigs at the national level in Canada. It is important to note that only one of these working groups, the Canadian Invasive Wild Pigs working group, reports directly into the ASF EMB. This group is developing a national wild pig strategy for Canada that will encompass work already completed at the regional and provincial levels.

These organization charts were made to be flexible as working groups and sub-working groups are constantly changing. There was a need to build in scheduled reviews and updating to these documents and to dedicate resources for this task. It is important that those involved in ASF planning and preparedness, but aren't directly members of the ASF EMB, be kept up to date with a knowledge of who is doing what. This proved to be a very successful strategy and makes others feel involved in the overall process.

3.4 The importance of regular status updates

An important role of the ASF coordination team was to regularly check-in on the status of each working group and sub-working group. The decision was made early to integrate a member of our team into each ASF EMB working group in order to keep up to date and ensure that regular progress reports occurred to the ASF EMB. The most successful strategy implemented for these routine updates was for the working group to bring their recommendations (instead of a list of questions) to the ASF EMB for discussion and feedback. This allowed the work of each group to be highlighted at the ASF EMB and provided direction for future tasks. These updates helped ASF EMB members understand the status of each working group and what to expect from them in the future. ASF EMB meetings are scheduled to occur on the same day, every second week, for 1.5 hours duration. Meeting agendas are populated by a small sub-committee of members from the NFAHWC ASF coordination team, Animal Health Emergency Management (AHEM) and CFIA that meet once per week for this purpose. Since ASF planning and preparedness is an ongoing effort, status updates for each working group and accomplishments will be highlighted in the oral presentation that will be given at the Banff Pork Seminar in January 2022.

4. Conclusions

Everyone has a role in preventing ASF from entering Canada. This starts by communicating the following to our peers and family members:

- Do not bring back pork or pork products from infected countries through travel or by mail
- Treat biosecurity seriously! On-farm biosecurity practices are key to preventing ASF and other diseases from infecting pigs
- If you have a pig do not feed it kitchen waste or food scraps. Reminder that this practice is prohibited in Canada.
- If you own pigs, contact a veterinarian immediately if disease is suspected and do not transport sick pigs to another location
- ASF can mimic other diseases commonly found on Canadian swine farms. Let a veterinarian determine the cause of disease and implement measures that will prevent the spread of disease

Time is a gift that has provided us the ability to pre-plan for a foreign animal disease in advance of detection. It takes time to learn how federal, provincial, territorial and industry (FPTI) partners can work together collaboratively towards the same goal. This structure should serve us well when needed for response efforts. Other agricultural species groups should be encouraged to replicate the ASF EMB model and implement the successful strategies outlined above. If ASF is not detected in Canada, this work can still be used for many different diseases or emergency response scenarios. So many have accomplished so much on ASF planning and preparedness. We will continue to implement the successful strategies outlined above. We will also continue to address any identified planning gaps in a timely manor and ensure that best communication practices amongst all involved continue.

Views from an Agricultural Immigrant

Vincent ter Beek

People often ask me how come I ended up in the swine business? After all, in college I read History and Journalism – topics quite unrelated to animal husbandry. Having obtained my MA, I pursued my studies in the United Kingdom to do a post-graduate newspaper journalism course in English. I've been fascinated by the English language since I was young, so it sounded like a logical choice to make sure I could do my job not only in Dutch, but also in English.

Having worked for a while for two Dutch local dailies, I noticed an advert of the publishing house of *Pig Progress*. They were looking for a native English speaker who was able to write journalistically and who knew his way around in the swine business. I may not be a native, but I manage well with my English, so I thought I might tick two out of three boxes. I decided to apply. I got an invitation and during the interview I admitted that I had no idea about the swine business other than I had been on a pig farm once. And that I would be happy to learn. It worked, a few days later I got that phone call. Last December, that was 16 years ago.

The Agricultural Immigrant

Over these years I've grown completely accustomed to working in the world of Feed Conversion Rates, sentinel pigs, anti-nutritional factors, *Streptococcus suis* and Large Whites. Yet even though I know my way pretty well in the business, I have never felt a complete insider in the global swine industry.

I think one can compare it to being a geographical immigrant moving from the Netherlands into Canada. I could try and get accustomed to new culinary challenges, yet the food wouldn't taste like it would back home. The command of language might be perfectly understandable, it may *just* sound a touch funny. Even though I would know how to sing *O Canada*, the only national anthem to really bring goosebumps would be the Dutch one. And although I'd keep an open mind, I would always continue to prefer soccer over hockey.

It doesn't mean I have any negative sentiments towards the swine business – far from that – but I tend to view pig farms differently than some of the passionate visitors of the Banff Pork Seminar may do. I have spoken to pig farmers who have that feeling of "coming home" as soon as they step into a pig house. To me a pig farm smells like a pig farm, and that is absolutely fine. Similarly, I'm not the most exciting dinner table guest if people start talking about feed prices the entire evening. At some point I would start steering the conversation towards soccer, politics or – obviously – history.

It took me some while to realize that being the immigrant is the role that suits me perfectly as an international journalist. Had I grown up in agriculture, then I would not have been able to apply a professional distance which I feel is also necessary from time to time to do my job well. So it happens that my Twitter timeline is a curious amalgam of people living in cities and working in journalism (often left-wing and sometimes vocal vegans), and people working in agriculture (complaining that many people in cities have lost touch with reality). I listen, will treat everyone with respect, but it's in my best interest not to choose sides.

How it Started

It was a role that grew on me. In the beginning I tried to make sure that I would know every technical term like anybody else. When the people in the business would be throwing around abbreviations I wanted to know them all. C:F, ADG, MCFA, PRRS – I'd have to nod and if somehow somebody would mention something I didn't know, I would look it up immediately afterwards.

Getting to know my way into a new business, however, was a steep learning curve. I suppose that is the case with many businesses – and the swine business is no exception. Various things I had no idea about. For instance, at that one-off visit to the pig farm I wrote about earlier, I asked the producer how many pigs he had on-site, to be able to say something about the size of the farm for a newspaper article. In his response he murmured something about the number of sows and the number of piglets.

I recall thinking: "Why talking about – how does he call them – *sows*? Surely I asked about the number of *pigs*… Skip the jargon, stop making things more difficult than necessary!"

It took a few years before I truly understood why the pig producer had been responding the way he had.

Similarly, before I started working for *Pig Progress* I had no idea that it was common practice to castrate male pigs. Just one of the things that the entire swine business assumes to be completely normal, that nobody even gives it a thought that there could be customers who simply don't have a clue. Ever since I'm convinced that if the industry would stop castrating overnight and not tell anyone, the feared outcry may not be half as big as expected.

Ever since I have managed to get my head around many aspects of international swine production. I know my way around with management, biosecurity, housing and equipment, breeding, the veterinary world – yet undoubtedly one of the fields that I struggle with is the world of animal nutrition. Of all aspects of swine production, especially the nutritionist part I found a castle that is hardest to conquer. Obviously I know the basics, but there are so many ingredients, and parts of ingredients, additives, minerals, substitutes, classifications or nicknames of ingredients that oftentimes at an international swine meetings, hitherto unknown nutritional terms get smoothly dropped without further introduction, followed by the next, and a next, plus a series of assumptions deemed logical to any member of the audience except me, topped up with for me unintelligible graphs and tables, that in despair I find myself staring out of the window, hoping the next speaker will have something more understandable to share.

And yes, that still happens fairly regularly after 16 years.
I hate to share it – but in the context of my presentation, this is relevant. Because I feel it might perhaps also say something about the swine industry. If a benevolent

outsider, after 16 years, is not capable from time to time to grasp the message of what is being shared, then which conclusion to draw? Obviously, it means that I still have a lot to learn. But I think it also means – sorry to be blunt – that perhaps the swine business is also a place full of in-crowd people, of specialists understanding each other but never really care about how outsiders may perceive the business.

The Gap Between Agriculture and Society

Let me now turn to the famous gap that the swine business often points to. City people have lost touch with agriculture, I often hear. The National Pork Producers Council in the USA from time to time stated that they wanted to "educate" the city people. They've lived away from agricultural reality for too long, they think milk comes in bottles. They don't realise that to for a pork chop, a sausage or spare ribs to be created, a living creature has to be grown and slaughtered.

Now the logical question I have for the visitors of the Banff Pork Seminar is this: who is to blame for that gap between society and agriculture? And it is a rhetorical question as far as I'm concerned – there's "blame" on both sides of the gap. I hope that the words I wrote a few paragraphs back can show as an illustration that the swine business is not exactly super-straightforward to get to know. On the contrary, it takes perseverance.

And that is not surprising. Ever since the 1950s, the swine business everywhere in the world has been developing strongly. Rather than having multi-species farms, specialized farms were created everywhere. Some farms only do breeding, others only have finisher pigs. With farms fully equipped, in a safe remote place, and all the conditions met for perfect pig production. Decent ventilation; a great lay-out, sophisticated feeding systems, state-of-the-art troughs, next-level biosecurity protocols, extremely high-tech feed formulation, well designed and differentiated roads in and out, properly protected by rows of fences to keep out unwanted visitors, whether they are of animal or human nature.

The Instagram Test

Now what does that look like to outsiders? Imagine John and Anna are driving up all the way from Vancouver but alas, somewhere in Alberta their car broke down. They park at the roadside, and get out of their car.

"Do you smell what I smell? What *is* that?"
"I think it must be that building over there. I think they keep animals in there. Goats. Or cows or so."
"I think it's pigs."
"If you say so." [Stares] "If they keep pigs over there, then why don't we see then?"
"No idea. It certainly looks more like a prison with all that wire."

It's a credible way a conversation can go. And yes, they could do with a bit of additional education. In the Netherlands, quite a few farms have therefore added a viewing gallery to their pig houses. Locals on bicycle tours – or people like John and Anna – are welcomed to step in, and take a look through the windows in a special guest corridor, and can see what is going on inside, together with signs explaining what is happening.

That is one step. But I think something else also needs to happen. For opening up to the outside world, and become more understood, it is also important to ask the question: does the swine business want to show everything that is happening in the farm? For that, I like to think of the Instagram test. If a pig producer would want to be 100% transparent and share and "educate" everything with followers, would he be able to share pictures of farrowing crates and get it explained in such a way that city people will understand and appreciate the fact that an animal cannot move for a month? And how about tail docking? Castration?

Painful questions to ask perhaps – and I'd like to remind the readers I'm not asking them for activist reasons. They are relevant because this will form the license to produce in the near future. We all know that by 2050 there will be over 9 billion mouths to feed. They may not all be eating meat, but if everyone at least has some money to spend the demand to meat will only grow. And where will these extra people of the future live? Most likely the vast majority will live in cities. Which means that the number of people detached from agriculture will exponentially rise. Which also means that the scrutiny towards production methods is also likely to grow. The Instagram test therefore is not hypothetical – for swine producers it holds the key to future production, because increasingly people will ask how their food was made.

In the European Union they are currently talking about getting rid of farrowing crates by 2027 – and I am convinced technology will find new answers to the old problems of overlaying and aggressive sows. Tail docking as well as castration are also practices which are being debated and even though at the moment no clear steps are being taken in terms of legislation, I have no doubt that these practices will disappear.

The swine business can of course choose to ignore these sounds, dismiss them as the next passer-by having dropped some idiot thoughts or go back to their day-to-day job. Yet I'm not a passer-by. As an agricultural immigrant, I've stood for 16 years on the bridge spanning the gap between society and agriculture.

And today I've done what I was asked to – I shared my view from that bridge.

Ontario Pork

Celebrating 75 years of progress and resilience

Water Quantity and Quality On-Farm Assessment

Susan Watkins, Emeritus Professor

University of Arkansas System's Division of Agriculture
Cooperative Extension Service
Fayetteville, AR
Email: swatkin@uark.edu

Much has been learned about the contribution of water supplies to unsatisfactory growth rates, feed conversions, livability and reproduction in poultry and swine. The following checklist outlines a process for evaluating both quality and quantity when trouble shooting or creating a preventative maintenance program for drinking water supplies with ROI considerations. Water is the single greatest input and unfortunately has not always received its due recognition for how it can impact performance. Good news, swine industry is recognizing the positive results the poultry industry has achieved through the development of sound water programs and are now focusing on now to create their own programs.

The following can be used as a guideline for planning and implementing a methodical water system inspection process that provides producers with insight into this vulnerable part of production operations. Producers and support personnel can use the knowledge gained from this process to assess systems, identify weak points and create a monitoring program. By having sound data collected from the water monitoring program, then accurate correlations can be conducted with production and economic data.

Inspect System

Identify Source-Typically the sources fall into one of three categories: well, municipal, surface, other

- Exam well or surface water design
- Determine if any unusual conditions have occurred such as flooding or droughts. Does the area have activities like row-crop farming, mining or fracking?
- Check for well-head damage and determine if any maintenance has been done on the system such as pump replacement.
- Measure gallons per minute (gpm) at source. Rainwave flow meter can be attached to hydrants, but verify water hose draining from Rainwave does not restrict the flow during test
- What water treatments are used at the source- and in what order?
- Collect sterile water sample as close to source as possible for total bacteria, coliforms and mineral and pH analysis
- Visually examine water, smell and taste it

Storage

- Size/holding capacity/volume, material, color, location, cleanliness
- How does water flow into and out of storage?
- Does water leave storage directly from bottom or is it elevated off bottom?
- How long does water remain in storage? What is dwell time in storage between flocks and during flock?

- How often is storage cleaned and can it be completely cleaned including bottom? If cleaned, how and with what?
- Is water supply treated prior to or after storage?
- What products or treatments are used before and after storage?
- If concerns about water quality changing during storage, collect sterile water sample post storage for total bacteria and mineral and pH

Distribution

- Verify pipe size(s) and type of plumbing material (pvc, metal) from source to birds.
- Determine if underground water lines can be cleaned from source to barns.
- Have lines been cleaned and with what?

Barns-Entry Way/Ante Room

- Is there daily water usage information for flocks? Are there consumption patterns throughout the day?
- Check pressure reducer-maintenance schedule, has screen been checked for sediment build-up, are there pressure gauges pre and post
- Inspect filters –are they clean, how often are they changed? Do they clog frequently? (If so, submit filter covered in clogging material to Watkins Water Lab for mineral/microbial analysis)
- What is gpm flow at barn entrance

- Inspect injectors-what is gpm rating, do connector hoses create flow restrictions, Have injectors been repaired? Has injection accuracy been verified? How many injectors? What are they used for? Are they cleaned between uses: What is container used to hold injected products? Is it sealed, used for anything else?
- What products are added? If sanitation products/acidification is used, is there time for products to mix into water prior to additional injections? How often are products added?
- Is there any verification of sanitizer residual in water?

Drinker System

- Verify system type
- Verify age and length of each water line and drinker spacing and # per line
- Determine #birds/drinker-during brooding and grow-out
- Inspect hoses and connectors, with particular attention to drop hoses into water lines
- Regulators-age, maintenance, chemicals used (cleaners, sanitizers, acids)
- Have water lines been cut open? If so, what was observed on inside of line?
- Do lines experience sediment buildup? Do drinkers leak due to sediment build-up?
- Has system been cleaned? When and with what?

- Are water lines flushed? How often?
- Inspect water line drain hoses, do they allow water flow to move freely or are there restriction points?
- Is there any documentation of water sanitation product residual in drinker lines in front and back of barn
- Are all water lines used? How many drinkers per animal?
- Visually examine water from water line, smell and taste it
- Collect sterile drip water samples from water line but do so in a way that prevents contamination of sample from fans and moving air
- Collect swab samples from end of water lines and when possible, from inside regulator, also prevent contamination from air from fans, etc.

Cool Cells and Foggers

- Are any products used?
- Inspect system and if challenges with cool cells-turning to mush or mineral build-up clogging air flow, take sample of water from cool cell recirculation tank for mineral and pH analysis, particularly after cool cells have been used for several days
- How often are fogger lines used?
- Are they cleaned during non-use or prior to warm weather?

Water Analysis to Request

Standard Analysis: Analyze source samples for minerals and pH, total bacteria and coliforms. Test drip and swab samples from water lines for total bacteria.

Carefully collect samples to prevent contamination from hands, air or outside of water line. If using a different lab from ones listed below, confirm lab will analyze for: Calcium, magnesium, iron, manganese, sulfur (sulfates), chloride, sodium, nitrates, nitrites and pH.

Additional tests should be requested if the following symptoms or concerns are present.

- If a clear to white filmy, stringy slime is present then request fungus, yeast and mold and general microbial profile. Collect the slime in as sterile as possible manner with swab
- Pinkish, reddish, brown slime in storage tanks, water filters or lines is present then request pseudomonas analysis
- Concerns that treatments are not working properly, ie filtration, Reverse Osmosis, magnets- submit a sample collected post treatment for mineral and pH analysis
- If animals are experiencing health challenges such as respiratory or dysentery infections then collect drip samples from near end of water lines and swab samples from within nipple pipes and end of water lines. Request total microbial profile including staphylococcus as well as total bacteria, yeast and mold analysis Ask if diagnostic labs can conduct a microbial profile on samples.
- If water products such as therapeutics or organic/inorganic acids are frequently used, then request bacteria, yeast and mold analysis

Collect samples in as sterile as possible manner for microbial analysis and store in a cooler with ice packs until submitted to lab. For best results, submit sample within 24 hours of collection but within 4 days is acceptable as long as samples are kept cold but not frozen.

BUILD
ON SOLID
FOOTING.
barkman
THE BEST SLATS FOR YOUR HOG BARN
Barkman offers a wide selection of slat styles and sizes featuring safer footing, improved sanitation and long-term durability.
BarkmanConcrete.com | 800 461 2278

FEEDING SYSTEMS
GROUP GESTATION
Gestal
3G
• Simple management of free gestating sows
• Individually feed your sows with precision
LACTATION
Optimize feed intake and management in lactation.
QUATTRO
FOCUS
SOLO
JYGA
Technologies
High Performance Made Easy
jygatech.com • 1 866 333-7853

Identifying What's in Your Water and How to Improve It

Martine Giguère, agr. MBA

Agro-Bio Contrôle inc.
Email: martine@agrobiocontrole.ca

The importance of water quality in farming is often underestimated. While the nutritional value of food is often analyzed in detail, water is often overlooked. Providing animals with good-quality water can help reduce the risks of health problems and improve animal performance. In addition to ensuring they receive enough water, its quality must be checked. This is done by diagnosing, cleaning and disinfecting the entire water distribution system and continuing to purify the water. Adjusting pH using organic acids is highly advantageous in pig production, especially among piglets.

Establishing a Water-Quality Diagnosis

Performing a physicochemical and bacteriological analysis helps paint a picture of the water and is the starting point for establishing a diagnosis. Certain basic parameters are good indicators of water quality; in particular, pH, hardness, total dissolved solids, and the content of various minerals such as iron, manganese, sulphate, nitrates, nitrites, and so forth. The analysis is then interpreted. Depending on the type of farming, some parameters can be guided by standards; for example, iron content for calf farms, while other parameters require a more advanced knowledge of water chemistry. These parameters should also include knowing if the water source is contaminated with bacteria such as *E. coli* as well as fecal and total coliforms. The tolerance and standards defining the presence of bacteria become better known, and the bacteria in water is somewhat or not at all tolerated.

Increasingly, on-farm food-safety programs are making more demands on water quality. Several programs require an annual sampling and analysis of certain parameters, such as the Dairy Farmers of Canada's ProAction program, which requires a bacteriological analysis every year (see Table 1). The Canadian Pork Excellence (CPE) program (module 4)[1] requires an annual water analysis be carried out on non-municipal water (see Tableau 2) for nitrates or total coliforms. High nitrate levels can be an indicator of bacteriological contamination. This is why the CPE program recommends a bacteriological analysis when the nitrate content is high. Even though it is known that the presence of fecal coliforms (standard of <1 CFU / ml) can cause diarrhea in piglets, there is no requirement to systematically test for fecal coliforms.

Once the results of the analysis are received, it is important to prioritize the actions to be taken. The factors affecting water quality are not always complex or expensive to regulate, such as pH level and controlling certain bacteria and biofilm (see box entitled Biofilm).

Table 1. Water-quality requirements – ProAction

Bacteriological parameters	Standards
Atypical bacteria	200 CFU/100 ml
E. coli	0 CFU/100 ml
Enterococci	0 CFU/100 ml
Total coliforms	10 CFU/100 ml

Table 2. Program water-quality requirements – Canadian Pork Excellence

Parameters	Standards
Nitrates	300 mg/100 ml
Total coliforms	10 CFU/100 ml

Maintaining Water Quality

Even if the water meets all the requirements at the entry point, contamination is often observed as it simply circulates through the storage tanks and distribution system (water lines) to the pipette/water trough. The natural formation of biofilm (see box) contributes to this contamination. To maintain water quality, the first step is to clean and disinfect the entire distribution system. Continuous water purification from then on will sustain its quality and prevent biofilm from forming. Chlorine or hydrogen peroxide injections are good ways to sanitize the water. Which one to use depends on the physicochemical parameters of the water, in particular pH, but also on the presence of bacteria. Among other factors, the effectiveness of chlorine diminishes as pH increases. Various water-purification options will be proposed in the presentation.

Adjusting pH with an acidifier based on organic acids is also advantageous (see Table 3) in pig production, in particular among piglets. Water acidification at weaning (0–21 days) brings benefits both in terms of health and animal performance. In the nursery, we recommend keeping the water pH at 3.8–4 for the first 0–21 days. Average daily gain can therefore be increased from 10–12 grams per day while improving daily consumption and significantly reducing diarrhea (see Table 4). For the rest of the rearing period, the pH of the water is kept at 6.

Table 3. Benefits of organic acids

Improves protein digestion – activation of digestive enzyme proteins (pepsinogen to pepsin)
Regulates the microflora in the intestine – support diversity of microbial balance
Support intestinal health integrity

Table 4. Reduce the Scour Incidence in Nurseries – Water acidification at pH 4 (0–21 days) – initial pH 8.8

	Control	Polyacid	P Value
ADG 0–3 weeks (g/kg)	345	357	0.30
ADG global (g/kg)	487	488	0.90
Global scour index	**0.18**[b]	**0.11**[a]	**0.05**
In nurseries (0–21 days), water acidification with Polyacid, a blend of organic and inorganic acids, at pH 4 significantly reduced diarrhea.			

Source: Groupe Cérès, 2016

Good water management will be beneficial for animal health and performance. However, to choose the right solution, a proper diagnosis must be established. Before choosing a water treatment, make sure you have pinpointed the problem.

Cleaning and Disinfecting Water Lines

Once the water diagnosis is done, the water distribution system must be cleaned and disinfected. Even when the water entering the barn is free of bacteria, contamination has been observed as it circulates throughout the distribution system and to the animal's drinking trough. There are two ways to check water quality in the distribution system. First, a bacteriological analysis of water samples from the tank and from the watering trough at the back of the barn could reveal the presence of bacteria. This method is less precise than analyzing the biofilm in the water lines. Biofilm analysis is more expensive and requires a little planning as it involves accessing the water lines to take a sample using sponges. These sponges are preserved in culture media and must be kept cool. During sampling, the outside of the water line must be thoroughly disinfected before inserting the tongs and sponge (see the video from the Chaire en recherche avicole de l'Université de Montréal: https://www.cra-fmv.org/evaluation-des-pathogenes-biofilms). This type of analysis makes it possible to measure not only the total aerobic flora present in the water line but the specific bacteria as well. An interpretation guideline for total aerobic flora gives an overall idea of the state of the line:

Table 5. Total aerobic flora.

Total aerobic flora count (CFU/ml)
Results interpretation guidelines
< 1,000 CFU/ml → line ok 1,000 to 10,000 CFU/ml → line to watch > 10,000 CFU/ml → problematic line

Test results prove that cleaning with soap is an important step that must be carries out. In fact, the total aerobic flora count is much lower when cleaned with a chlorinated alkaline soap and then disinfected with a product based on peracetic acid (see Table 6).

Table 6: Summary of the biofilm analysis results – Poultry house

Level	Treatment	Sampling	Assessment of total aerobic flora (CFU/ml)	Results interpretation
1st floor	None	Pre-treatment	6,500	1,000 to 10,000 CFU/ml → line to watch
	Cold water + detergent	Post-detergent	10	< 1,000 CFU/ml → line ok
	Oxylis 2% – 24 h	Post-disinfection	150	< 1,000 CFU/ml → line ok
2nd floor	Hot water + detergent	Post-detergent	610	< 1,000 CFU/ml → line ok

	Oxylis 2% – 24 h	Post-disinfection	10	< 1,000 CFU/ml → line ok
3rd floor	None	Pre-treatment	17,200	> 10,000 CFU/ml → problematic line
	Oxylis 2% – 24 h	Post-disinfection	1,230	1,000 to 10,000 CFU/ml → line to watch
	Oxylis 2% – 24 h + Aquaflush	Post-disinfection	2,790	1,000 to 10,000 CFU/ml → line to watch

In another trial[2], the impact of the water-line cleaning with a peracetic acid, water disinfection with chlorine and water acidification with a blend of inorganic and organic acid was compared with the water line not being cleaned. Weight gain from weaning to week 6 was measured. It was observed that average daily gain increased by 21% when the line was cleaned, and the water disinfected and acidified, compared with control. The feed efficiency improved by 13% compared with control, and a difference of 2.49 kg was observed in the piglet after 6 weeks. So, water-line cleaning, water disinfection and acidification help improve animal performance (see Figure 1).

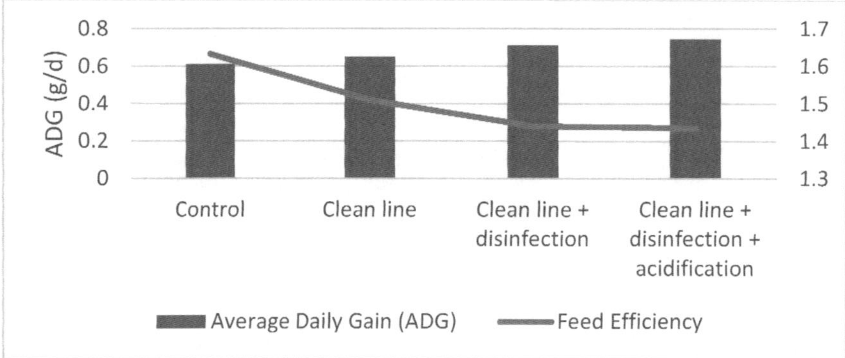

Figure 1. Water quality and animal performance in nursery.

Source: Examining the impact of water line cleaning, disinfection and pH reduction on nursery pigs, Andrea Pitkin, DVM, MS, Fairmont Veterinary Clinic, 2012 AASV, Denver, CO

Setting Up an Intervention Plan

Establishing a diagnosis is essential for improving water quality and animal performance. Taking a sample is a relatively simple way to obtain an overall picture. Cleaning and disinfecting the water lines must be thoroughly planned for successful results. A work plan must be developed, buildings divided into sub-sections, and so forth. Once the water lines have been thoroughly cleaned, continuous water purification is paramount to preventing recontamination. Cleaning and disinfecting can even be carried out while the animals are in the barn, as long as the steps are planned out well.

Microorganisms will adhere to the interior walls of the pipes. They then form a matrix in which microorganisms naturally accumulate. The biofilm becomes a favourable place for bacteria to multiply. As water flows thought the pipes, biofilm portions fall off and contaminate the drinking water. The quantity of biofilms varies according to the presence of bacteria, concentration of nutrients (water hardness and minerals), type of materials, and so forth. In some cases, the presence of biofilm considerably reduces water flow, thus reducing the animals' water consumption.

References

[1]https://www.cpc-ccp.com/uploads/userfiles/files/2018%20CPE%20PigSAFE%20PigCARE%20Producer%20Manual%20WEB%20nov%209.pdf

[2]Examining the impact of water line cleaning, disinfection and pH reduction on nursery pigs, Andrea Pitkin, DVM, MS, Fairmont Veterinary Clinic, 2012 AASV, Denver, CO

MAPLE LEAF
is delighted to participate in the
2022
Banff Pork Seminar
as a
Premier Plus Sponsor
For information regarding Marketing, Pricing, Settlement, Logistics and Contracting, Call
1-877-937-4647 Brandon (toll-free)
*Produced in Western Canada
Canada's leading producer of premium quality, value-added fresh and further processed meat and meals products.

Precision Nutrition and Feeding Enabled by Technology

B.R. Frederick, Ph.D.

Cargill Animal Nutrition, Lewisburg, Ohio

1. Introduction

Accurate and precision feeding is a rather simple concept to comprehend in theory: meet the pig's nutrient requirements without exceeding on a consistent basis. This statement includes the accuracy of knowing the pig's requirements while precisely adjusting to their changing needs. One without the other would lead to precisely inaccurate nutrition. However, execution of precision nutrition and feeding is easier said than done for several reasons. Typically, we base our feeding strategies on populations of pigs not the individual and our equipment is designed to feed groups of animals rather than just one. Our production practices, industry technology, and scientific advancements have resulted in an efficient food system. However, further technology advancements will be required to further improve our efficiency and ability to feed a growing population.

How might one go about increasing precision, you might ask? One of Wayne Cast's many sayings is "we formulate with a caliper, our feed mills mark it with chalk, and we deliver feed with an axe". This quote refers the precision, perceived or real, to three decimal places within formulation moving through the implementation and delivery process subsequently leading to actual nutrient amount consumed through a day within a specific growing period. Each step in the process has a chance to reduce precision and accuracy.

There may well be a time in the future that each pig receives a different diet every day that most closely matches individual requirements. However, this scenario which would ultimately result in applicable, profitable production is well into the future. This paper and subsequent presentation will focus on some fundamental and applicable implementation, as well as a vision that can be realized in the next 5 to 10+ years.

2. Ingredient Valuation

Regardless of the phase of production a strong foundation within ingredient valuation is essential. Proper ingredient evaluation is a cornerstone of improving precision. Confidence in nutrient loadings will lead to less safety margins when setting nutrient specifications in formulation (Patience, 1996). Wet chemistry and analytics are typically the preferred method to obtain accurate nutrient values. However, this method is also the most timely and expensive. A well calibrated near infrared spectroscopy (NIR) can provide very good precision and increase the speed of analysis while reducing overall analytical costs (van Barneveld et al, 1999).

Rapid and accurate analysis of incoming ingredients is essential to achieving precision. The speed of analysis and accurate reflection of inbound ingredients is the most important factor. NIR technology continues to advance and provide more real time information. Instruments have moved from laboratory to ingredient receiving and are now inline at the point of ingredients going into the mixer at the feed mill. Each step closer to real time data provides an opportunity to improve precision and accuracy. Theoretically, inline NIR can yield the most real time data available. Thus, it could be possible that each batch of feed within the mixer is a different formulation. For large, high volume feed mills this concept is very interesting and becoming reality today.

3. Formulation, Diet Strategy and Ingredient Selection

Formulation, diet strategy and ingredient selection are all interdependent to ultimately determine the most profitable solution in any given market condition and producer objective. Given a strong sense of confidence with the accuracy and precision of nutrient loadings, the nutritionist can spend more time focusing on setting nutrient specifications to match the pig's requirement. Market conditions may drive producers to set performance goals that are below maximum potential and formulate diets accordingly

Once the diet strategy is set for market conditions, sourcing ingredients should not be overlooked. Selecting the proper ingredient sources based on nutrient values can result in $0.50 to $1.00 per pig in profit.

4. Feed Manufacturing and Delivery

Feed milling and manufacturing should not be overlooked as a critical step in precision nutrition. Ensuring consistent particle size reduction of major ingredients is important in delivering the proper nutrient specifications. A critical step in precision and accuracy is the inclusion of the proper amount of micro ingredients. For example, even moderate variation in the addition of synthetic amino acids can result in substantial impacts to diet cost by over supplementation. Even more costly is the impact of pig performance from under supplementation, which is more economically impactful given the investment in all other nutrients.

Timely and accurate feed delivery is important. Advancements in delivery technology have yielded options to ensure the right feed gets to the right pig at the right time. Feed bin technology can automatically monitor inventory to ensure pigs never run out of feed and help verify deliveries as expected. Various technology solutions exist to verify proper delivery to feed bins ranging from bar code validation, geofencing, and integrated systems on the feed truck themselves.

5. Application of Precision to the Pig

Animal agriculture in general is starting to see precision nutrition and feeding more common in other species. Individual, specialized feeding is notable in the Dairy industry. Milk output and quality are rapid feedback mechanisms that allows for adjustments to be made daily on an individual basis. Layers also provide rapid feedback on output based on egg quantity and quality, although may not be individual information in open rearing situations. Pigs on the other hand do not have the rapid output feedback compared to the two previous species. Individual records on sows and advancements in electronic feeding equipment are now unlocking possibilities of individual feeding even in an open pen environment.

Perhaps the most notable current advancements that are being explored and needed to further advance precision feeding is at the barn level. The sow farm is a prime candidate for precision feeding given the individual animal records, high investment cost per animal, and potential impact in terms of varying requirements through the breeding, gestation, transition, and lactation. The sows varying nutrient requirement, linked with potential transgenerational nutrition and epigenetics, provides exciting new areas to explore in the future. Equipment has and continues to develop to deliver the opportunity to capture these potentials. In a recent podcast Jeff Knott (Knott, 2021) outlined the potential areas to explore enabled by these technologies.

Another area that is deserving of more technology application is wean to market. First and foremost is to leverage current technology associated with barn environment. Greater adoption of "smart", connected controllers is a major advancement not only to the pig environment monitoring and intervention, but also powerful information to study potential nutrition by environment interactions to optimize performance. These advancements are followed closely by water monitoring and feed bin technology which yield valuable information for more real time leading indicators for a feedback loop to precision nutrition and potential modifications on diet strategy.

Feed intake monitoring equipment, primarily bin monitors, is becoming increasingly accessible and reliable. This leading indicator is important to help ensure pigs are consuming the proper amount of overall nutrients as well as provide the opportunity to intervene and more precisely feed the pig within the growth phase rather than waiting for a closeout. Precise feed intake linked to diet formulation can also be used to predict pig weight and optimum time to market. Not only is one able to more precisely feed and predict performance, but also fine tune marketing.

The next evolution in precision feeding is indented to validate that the feed disappearing from the bin is converting to pork. Machine vision have been developed to predict bodyweight to within about 3% of actual weight. This can further add information to the precision feeding concept. Most current installations have a few cameras per group or barn to predict average weights. In the future cost will likely make monitoring applicable for every pen enabling feeding by pen an option from an information perspective. Feeding systems would need to follow and be practical to execute to the pen level. There may be a time in the future that individual pigs are fed uniquely to their needs. However, that practical hurdle is a large one.

6. Conclusion

Accurate and precision nutrition starts with the foundation of ingredient valuation, proper formulation, feed manufacturing, and delivery. Additional technology will continue to come to market but durability, connectivity, and cost will dictate the rate of adoption. Companies and systems that connect relevant and valuable data together to make agile and perhaps automated decisions will have an advantage over those that do not.

References

Knott, J. (2021). Applications of Precision Nutrition. Gestal Swine Summit. https://jygatech.com/2021-gestal-swine-summit/.

Patience, J. F. (1996). Precision in swine feeding programs: an integrated approach. *Anim. Feed Sci. Technol.* 59 (1–3), pp.137-145.

van Barneveld RJ, Nuttall J, Flinn PC, Osborne B. (1999). NIR reflectance measurement of the digestible energy content of cereals for growing pigs. Journal of Near Infrared Spectroscopy 7:pp.1–7.

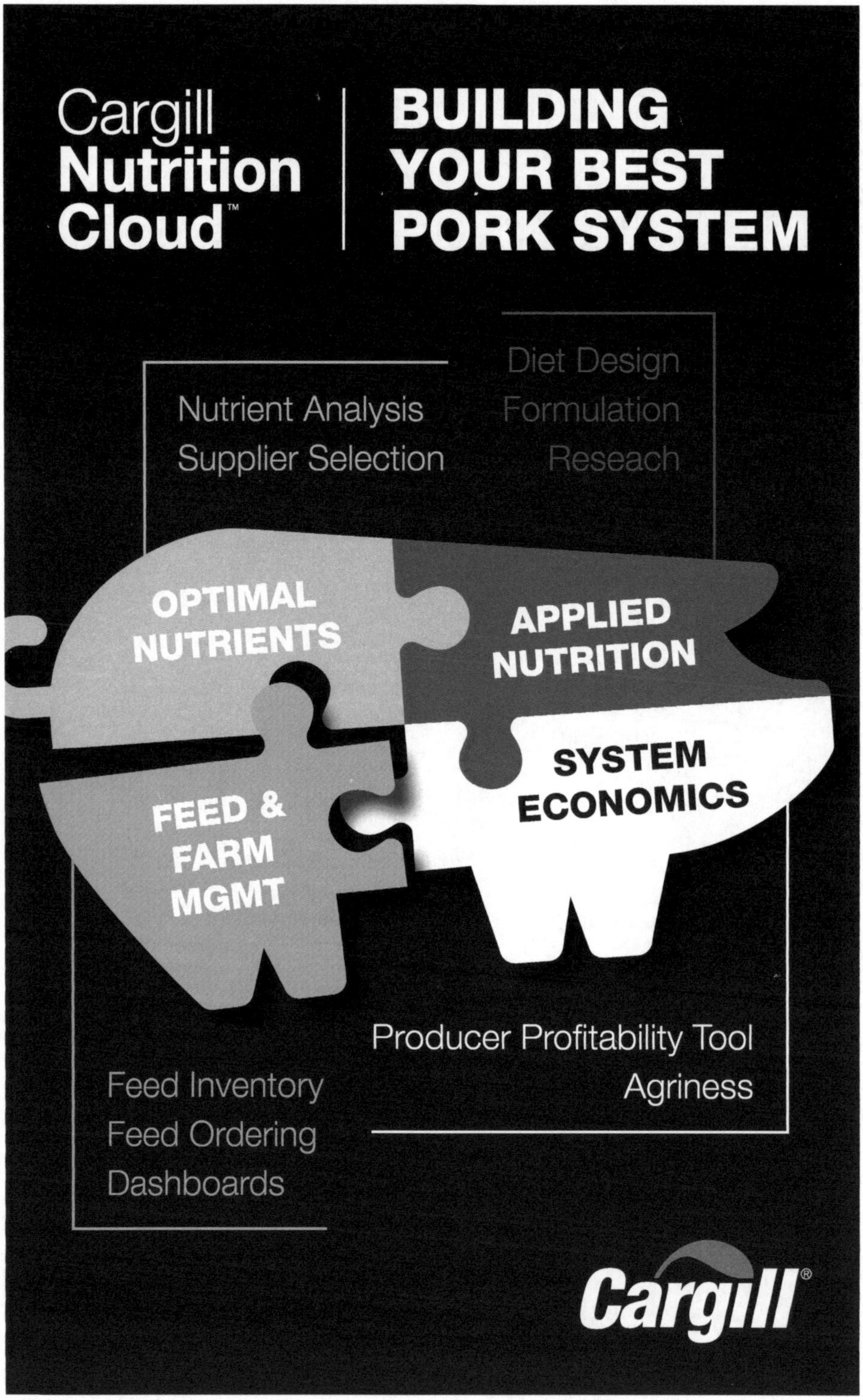
Cargill
Nutrition
Cloud™
BUILDING
YOUR BEST
PORK SYSTEM
Nutrient Analysis
Supplier Selection
Diet Design
Formulation
Reseach
OPTIMAL
NUTRIENTS
APPLIED
NUTRITION
SYSTEM
ECONOMICS
FEED &
FARM
MGMT
Producer Profitability Tool
Agriness
Feed Inventory
Feed Ordering
Dashboards
Cargill®

Precision Nutrition and Feeding Enabled by Technology

B.R. Frederick, Ph.D.

Cargill Animal Nutrition, Lewisburg, Ohio

1. Introduction

Accurate and precision feeding is a rather simple concept to comprehend in theory: meet the pig's nutrient requirements without exceeding on a consistent basis. This statement includes the accuracy of knowing the pig's requirements while precisely adjusting to their changing needs. One without the other would lead to precisely inaccurate nutrition. However, execution of precision nutrition and feeding is easier said than done for several reasons. Typically, we base our feeding strategies on populations of pigs not the individual and our equipment is designed to feed groups of animals rather than just one. Our production practices, industry technology, and scientific advancements have resulted in an efficient food system. However, further technology advancements will be required to further improve our efficiency and ability to feed a growing population.

How might one go about increasing precision, you might ask? One of Wayne Cast's many sayings is "we formulate with a caliper, our feed mills mark it with chalk, and we deliver feed with an axe". This quote refers the precision, perceived or real, to three decimal places within formulation moving through the implementation and delivery process subsequently leading to actual nutrient amount consumed through a day within a specific growing period. Each step in the process has a chance to reduce precision and accuracy.

There may well be a time in the future that each pig receives a different diet every day that most closely matches individual requirements. However, this scenario which would ultimately result in applicable, profitable production is well into the future. This paper and subsequent presentation will focus on some fundamental and applicable implementation, as well as a vision that can be realized in the next 5 to 10+ years.

2. Ingredient Valuation

Regardless of the phase of production a strong foundation within ingredient valuation is essential. Proper ingredient evaluation is a cornerstone of improving precision. Confidence in nutrient loadings will lead to less safety margins when setting nutrient specifications in formulation (Patience, 1996). Wet chemistry and analytics are typically the preferred method to obtain accurate nutrient values. However, this method is also the most timely and expensive. A well calibrated near infrared spectroscopy (NIR) can provide very good precision and increase the speed of analysis while reducing overall analytical costs (van Barneveld et al, 1999).

Rapid and accurate analysis of incoming ingredients is essential to achieving precision. The speed of analysis and accurate reflection of inbound ingredients is the most important factor. NIR technology continues to advance and provide more real time information. Instruments have moved from laboratory to ingredient receiving and are now inline at the point of ingredients going into the mixer at the feed mill. Each step closer to real time data provides an opportunity to improve precision and accuracy. Theoretically, inline NIR can yield the most real time data available. Thus, it could be possible that each batch of feed within the mixer is a different formulation. For large, high volume feed mills this concept is very interesting and becoming reality today.

3. Formulation, Diet Strategy and Ingredient Selection

Formulation, diet strategy and ingredient selection are all interdependent to ultimately determine the most profitable solution in any given market condition and producer objective. Given a strong sense of confidence with the accuracy and precision of nutrient loadings, the nutritionist can spend more time focusing on setting nutrient specifications to match the pig's requirement. Market conditions may drive producers to set performance goals that are below maximum potential and formulate diets accordingly

Once the diet strategy is set for market conditions, sourcing ingredients should not be overlooked. Selecting the proper ingredient sources based on nutrient values can result in \$0.50 to \$1.00 per pig in profit.

4. Feed Manufacturing and Delivery

Feed milling and manufacturing should not be overlooked as a critical step in precision nutrition. Ensuring consistent particle size reduction of major ingredients is important in delivering the proper nutrient specifications. A critical step in precision and accuracy is the inclusion of the proper amount of micro ingredients. For example, even moderate variation in the addition of synthetic amino acids can result in substantial impacts to diet cost by over supplementation. Even more costly is the impact of pig performance from under supplementation, which is more economically impactful given the investment in all other nutrients.

Timely and accurate feed delivery is important. Advancements in delivery technology have yielded options to ensure the right feed gets to the right pig at the right time. Feed bin technology can automatically monitor inventory to ensure pigs never run out of feed and help verify deliveries as expected. Various technology solutions exist to verify proper delivery to feed bins ranging from bar code validation, geofencing, and integrated systems on the feed truck themselves.

5. Application of Precision to the Pig

Animal agriculture in general is starting to see precision nutrition and feeding more common in other species. Individual, specialized feeding is notable in the Dairy industry. Milk output and quality are rapid feedback mechanisms that allows for adjustments to be made daily on an individual basis. Layers also provide rapid feedback on output based on egg quantity and quality, although may not be individual information in open rearing situations. Pigs on the other hand do not have the rapid output feedback compared to the two previous species. Individual records on sows and advancements in electronic feeding equipment are now unlocking possibilities of individual feeding even in an open pen environment.

Perhaps the most notable current advancements that are being explored and needed to further advance precision feeding is at the barn level. The sow farm is a prime candidate for precision feeding given the individual animal records, high investment cost per animal, and potential impact in terms of varying requirements through the breeding, gestation, transition, and lactation. The sows varying nutrient requirement, linked with potential transgenerational nutrition and epigenetics, provides exciting new areas to explore in the future. Equipment has and continues to develop to deliver the opportunity to capture these potentials. In a recent podcast Jeff Knott (Knott, 2021) outlined the potential areas to explore enabled by these technologies.

Another area that is deserving of more technology application is wean to market. First and foremost is to leverage current technology associated with barn environment. Greater adoption of “smart”, connected controllers is a major advancement not only to the pig environment monitoring and intervention, but also powerful information to study potential nutrition by environment interactions to optimize performance. These advancements are followed closely by water monitoring and feed bin technology which yield valuable information for more real time leading indicators for a feedback loop to precision nutrition and potential modifications on diet strategy.

Feed intake monitoring equipment, primarily bin monitors, is becoming increasingly accessible and reliable. This leading indicator is important to help ensure pigs are consuming the proper amount of overall nutrients as well as provide the opportunity to intervene and more precisely feed the pig within the growth phase rather than waiting for a closeout. Precise feed intake linked to diet formulation can also be used to predict pig weight and optimum time to market. Not only is one able to more precisely feed and predict performance, but also fine tune marketing.

The next evolution in precision feeding is indented to validate that the feed disappearing from the bin is converting to pork. Machine vision have been developed to predict bodyweight to within about 3% of actual weight. This can further add information to the precision feeding concept. Most current installations have a few cameras per group or barn to predict average weights. In the future cost will likely make monitoring applicable for every pen enabling feeding by pen an option from an information perspective. Feeding systems would need to follow and be practical to execute to the pen level. There may be a time in the future that individual pigs are fed uniquely to their needs. However, that practical hurdle is a large one.

6. Conclusion

Accurate and precision nutrition starts with the foundation of ingredient valuation, proper formulation, feed manufacturing, and delivery. Additional technology will continue to come to market but durability, connectivity, and cost will dictate the rate of adoption. Companies and systems that connect relevant and valuable data together to make agile and perhaps automated decisions will have an advantage over those that do not.

References

Knott, J. (2021). Applications of Precision Nutrition. Gestal Swine Summit. https://jygatech.com/2021-gestal-swine-summit/.

Patience, J. F. (1996). Precision in swine feeding programs: an integrated approach. *Anim. Feed Sci. Technol.* 59 (1–3), pp.137-145.

van Barneveld RJ, Nuttall J, Flinn PC, Osborne B. (1999). NIR reflectance measurement of the digestible energy content of cereals for growing pigs. Journal of Near Infrared Spectroscopy 7:pp.1–7.

Reducing Feed Cost in Grow-Finish: Alternative Feedstuffs and Production Strategies

Eduardo Beltranena[1], Miranda N. Smit[2], Malachy G. Young[3] and José L. Landero[3]

[1]University of Alberta, [2]Prairie Swine Centre Inc., [3]Gowans Feed Consulting

At time of writing, it seems unclear what 2022 will bring to pig producers. Lean Hog Futures had positive slopes increasing US$4/month to mid $90s from January to June 2021. Since, contract prices dived twice to ~$80 recovering to ~$90 for the first half of 2022. Canada - China relations continue to be our biggest unknown regarding pork and grain exports. Weekly hog margins were generous in Canada for the first 3 quarters of 2021, but most concerning has been the lack of summer rain. Total crop production is forecasted to decrease by 30%, production is estimated to have declined by 40% vs. 2020 and be 36% below the five-year average as drought affected 94% of agricultural land in Western Canada (Ag Canada). Grain prices will stay strong due to robust international demand, low carry in stocks, and limited fall grain harvested imply 25% decline in total supplies. Wheat yield this fall will be nearly one-third down vs. the 5-year average. Strong futures and low North American supply suggest CWRS prices will shut up >$400/tonne. The same tune plays for barley, expect 65% of the 5-year average yield. Oat exports are expected to reach record levels, with feed use forecasted to increase 8%. Given summer prices, we are already formulating corn grain and more soybean meal into hog diets. Expect Central corn and US soybean meal inclusions will increase given the low supply and high price of wheat, barley, and canola in the Prairies. Envisioning greater feed cost for hog producers in 2022, this paper reviews some strategies to mitigate feed cost including alternative ingredients and production strategies.

1. Feed Energy and Stocking Density

We are blessed in Canada to have cool barn temperatures at least two-thirds of the year. Most herds also are at a great distance from other barns and high biosecurity procedures are in place resulting in good health status allowing our hogs to reach high feed intake levels. We can therefore reduce feed energy density and thus diet cost and let hogs make up for the caloric difference by eating more. We have shown that this strategy can be profitable (Table 1.

Table 1. Effects of feeding diets with decreasing net energy (NE) level on hog profitability (Beltranena and Smit, 2015)

Mcal NE /kg	2.4	2.3	2.2	2.1
Diet cost, $/1000kg	249.51[a]	233.13[b]	216.22[c]	198.81[d]
Feed cost per kg gained, $	0.67[a]	0.63[b]	0.60[c]	0.57[d]
Feed cost per pig, $	62.50[a]	59.58[b]	56.72[c]	54.66[d]
Income per hog after subtracting feed cost, $	61.02[d]	63.50[c]	65.93[b]	71.43[a]

Decreasing net energy (NE) by 0.1 Mcal/kg linearly decreased feed cost by $17/tonne, feed cost per kg gained by 3¢, feed cost per hog by $2.60, and increased income subtracting feed cost by $3.50 per hog. Feeding 2.1 vs. 2.4 Mcal/kg resulted in >$10 greater profit per hog. But bankers hesitate lending money to producers for new barns and sows keep on being more prolific, so we have more hogs to stock growout pens. The question that came up was do hogs with limited pen space require more energy dense rations given limited feeder space or access to it. We asked 1,920 pigs (33 kg) housed in 96 fully slatted floor pens (6.1 × 2.4 m) equipped with 2 or 3 feeder spaces, and 18 or 22 barrows or gilts per pen, fed either low (2.2 Mcal/kg) or high (2.35 Mcal/kg) NE diets over 5 growth phases. For the entire trial, pigs fed low vs. high NE diets had 120 g/d greater feed disappearance but 20 g/g lower gain-to-feed (Table 2). Pens with 18 vs. 22 pigs had 60 g/d greater feed disappearance and 30 g/d greater weight gain. Pigs in pens with 3 vs. 2 feeding spaces had 50 g/d greater feed disappearance and 4 g/g lower gain-to feed. Feeding low vs. high NE diets reduced feed cost by $21.87/tonne, $3.34/hog, $0.03/kg gain, and increased income subtracting feed cost by $1.82/pig (Table 3). Housing 18 vs. 22 pigs per pen increased income subtracting feed cost by $1.98 per pig. Lack of interactions between feed energy level, feeder space, and group size for feed disappearance indicated that low net energy diets can be fed to hogs even if they have lower than recommended floor area allowance during part of the finishing phase.

Table 2. Effect of dietary net energy (NE) level (2.2 Mcal/kg [Low] vs. 2.35 Mcal/kg [High]), feeder spaces (2 vs. 3), and group size (18 vs. 22 pigs/pen) on overall (d 0 to 88) daily feed disappearance (ADF), daily weight gain (ADG), and feed efficiency (G:F) of growing-finishing barrows and gilts (Smit et al., 2021)

	NE level		Feeder space		Group size		P value		
	Low	High	2	3	18	22	NE level	Feeder space	Group size
ADF, kg	2.829	2.710	2.744	2.795	2.801	2.739	<0.001	<0.010	<0.001
ADG, kg	0.992	0.994	1.000	0.996	1.008	0.979	0.611	0.157	<0.001
G:F, kg:kg	0.351	0.368	0.361	0.357	0.361	0.358	<0.001	<0.050	0.126

Table 3. Effect of net energy (NE) level (2.2 Mcal/kg [Low] vs. 2.35 Mcal/kg [High]), feeder spaces (2 vs. 3) and group size (18 vs. 22 pigs/pen) on feed cost and income subtracting feed cost (ISFC; Smit et al., 2021)

	NE level		Feeder space		Group size		P value		
	Low	High	2	3	18	22	NE level	Feeder space	Group size
Feed cost/tonne	211.14	233.01	222.09	222.06	222.15	222.00	<0.001	0.951	0.754
Feed cost/pig	64.74	68.08	66.11	66.71	66.34	66.48	<0.001	0.148	0.742
Feed cost/kg gain	0.63	0.66	0.64	0.65	0.64	0.65	<0.001	0.202	0.052
ISFC	71.88	70.06	71.44	70.51	71.96	69.98	<0.050	0.206	<0.010

2. Hybrid Rye as Alternatives to Wheat Grain

Given high wheat prices for the last two years it would be unrealistic not to look for cereal alternatives. Wheat is more readily available, and one could even get food grade price for it, but if you grow your own either triticale or hybrid rye are worth considering yielding more pork per unit of land. New European, fall-planted hybrid rye grown in western Canada is more resistant to ergot and fusarium and has lower content of anti-nutritional factors than common rye. We evaluated the effect of feeding increasing hybrid rye level substituting wheat grain and non-starch polysaccharide (NSP) enzyme inclusion in diets fed to hogs. For the entire trial (d 0 to 76), pigs fed increasing hybrid rye level substituting wheat grain had decreased feed disappearance and weight gain (Table 4). Enzyme inclusion did not affect feed disappearance but tended to increase weight gain. Enzyme inclusion improved gain-to-feed ratio only in pigs fed the high rye level. Carcass traits were not affected by either increasing hybrid rye level substituting wheat grain or enzyme inclusion. Increasing dietary hybrid rye level substituting wheat grain increased cost per tonne of feed but did not affect feed cost per pig or per kg gain (Table 5). Enzyme inclusion increased cost per tonne of feed, but feed cost per pig and per kg gain were reduced. In conclusion, fall-planted hybrid rye can completely replace wheat grain in hog diets without affecting feed efficiency, feed cost per pig, or feed cost per kg gain. Inclusion of NSP enzyme would be recommended for diets including high rye levels to improve feed efficiency and weight gain.

Table 4. Daily feed disappearance (ADF), daily weight gain (ADG), and feed efficiency (G:F) of growing-finishing barrows and gilts (d 0 to 76) fed diets with increasing hybrid rye level substituting wheat grain with or without enzyme inclusion[1] (Smith et al., 2019).

	Rye substituting wheat			Enzyme		P-value		
	Low	Med	High	No	With	Rye	Enzyme	Sex
ADF kg/d	3.049[a]	2.975[a,b]	2.911[b]	2.966	2.990	0.0067	0.4664	<0.0001
ADG kg/d	1.013[a]	0.998[a,b]	0.971[b]	0.984	1.004	0.0114	0.0798	<0.0001
G:F kg/kg	0.333	0.336	0.334	0.332	0.337	0.7211	0.1148	<0.0001

[1]1200 mg/kg inclusion rate containing 1400 units of β-glucanase and 4500 units of xylanase per gram of product (Endofeed W DC, GNC Bioferm, Bradwell, SK, Canada).

Table 5. Feed cost and income subtracting feed cost (ISFC) of grower-finisher barrows and gilts fed diets with increasing hybrid rye level substituting wheat grain with or without enzyme inclusion (Smith et al., 2019).

	Rye substituting wheat			Enzyme		P value	
	Low	Medium	High	Without	With	Rye	Enzyme
Feed cost/tonne[4,5]	240.28[c]	241.28[b]	242.20[a]	240.36	242.15	<0.0001	<0.0001
Feed cost/pig[4]	80.57	81.55	81.75	82.14	80.44	0.3049	0.0142
Feed cost/kg weight gain[4]	0.94	0.95	0.95	0.96	0.94	0.3401	0.0129
ISFC[4]	30.66	28.20	28.31	28.53	29.59	0.1328	0.3382

3. Canola Meal or Soy Expeller to either Reduce or Increase Feed Energy

Feeding solvent-extracted canola meal (CM) with relatively high-fibre content or extruded-expelled soybean cake (SE) with remaining oil content offers the opportunity to either reduce or increase dietary net energy (NE) level at low cost. We fed 504 barrows and 504 gilts (33 kg) two NE levels (low 2.2 -2.25 Mcal/kg for grower and finisher, respectively] or high [2.37 – 2.44 Mcal/kg]) and two supplemental protein sources (CM [250-200 g/kg] or SE [150-125 g/kg]). For the overall trial (d 0 to 74), hogs fed low NE consumed 72 g/d more feed than those fed high NE diets, but their NE intake was 0.3 Mcal/d less (Table 6). Intake was 0.2 Mcal/d greater for pigs fed SE that grew 37 g/d faster than those fed CM. Dietary NE level did not affect weight gain. Gain-to-feed was 0.014 kg/kg greater for hogs fed high vs. low NE and 0.012 kg/kg greater for hogs fed SE vs.CM diets. Dressing was 6 g/kg greater feeding high vs. low NE or SE vs. CM diets. Carcass weight was 0.8 kg greater for hogs fed high vs. low NE; 1.4 kg greater for hogs fed SE vs. CM diets. Hogs fed SE averaged 2.3 mm greater loin depth than those fed CM. Dietary energy level or supplemental protein source had no effect on backfat depth, pork yield, or carcass index. Hogs fed CM stayed 2.6 d longer in the barn than those fed SE. Feeding CM reduced feed cost per hog shipped by $1.5 (Table 7). In conclusion, feeding reduced NE diets to hogs did not result in major changes in growth performance. Abruptly introducing 250 g/kg CM in the grower phase diets was a challenge to pigs. Pigs fed CM never caught up to those fed more palatable, lower fibre SE likely in part because we slightly overestimated the standardized amino acid content of CM and slightly underestimated the NE content of SE.

Table 6. Effect of dietary NE level (2.23 -2.24 Mcal/kg [low] or 2.37 – 2.45 Mcal/kg [high]), supplemental protein source (soy expeller [SE] or canola meal [CM]), and sex on overall (d 0 to 74) daily feed disappearance (ADF), weight gain (ADG), and feed efficiency (G:F) of growing-finishing barrows and gilts[a] (Smit et al., 2018).

	NE level		Protein		Sex		P value	
	Low	High	CM	SE	Barrows	Gilts	NE level	Protein
ADFI, kg[b]	2.785	2.713	2.743	2.755	2.935	2.564	<0.001	0.503
ADG, kg	1.002	1.011	0.988	1.025	1.047	0.966	0.164	<0.001
G:F	0.360	0.374	0.361	0.373	0.356	0.378	<0.001	<0.001

Table 7. Effect of dietary net energy (NE) level (2.23 -2.24 Mcal/kg [low] or 2.37 – 2.45 Mcal/kg [high]) and protein source (canola meal [CM] vs. soy expeller [SE]) on feed cost and gross income margin subtracting feed cost (ISFC; Smit et al., 2018).

	NE Level		Protein		*P* value	
	Low	High	CM	SE	NE	Protein
Feed cost/tonne	265.86	294.24	276.17	283.93	<0.001	<0.001
Feed cost/kg BW gain	0.80	0.84	0.82	0.81	<0.001	0.070
Feed cost/ shipped hog	78.54	82.87	79.96	81.46	<0.001	0.061
ISFC/shipped hog	68.84	66.09	67.46	67.78	<0.010	0.480

4. Faba Bean Replacing Soybean Meal

Pulse growers have been prioritizing production of zero-tannin, white-flowered cultivars (Snowbird, Snowdrop and Tabasco) as tannins reduce both starch and protein digestibility. However, tannin concentrated on the bean outer hull enhance frost tolerance of colour-flowered cultivars (Fabelle, Malik and Tabor). Furthermore, some of the tannin cultivars are low in vicine and covicine, compounds that cause favism (G6PDD) in humans, an inborn error predisposing red blood cells breakdown. Frost tolerance is a desired agronomic trait as faba bean takes ~2 weeks longer than field pea to reach maturity and there are differences among faba bean cultivars. Late planting due to a wet spring and(or) early frost around fall harvest time are unavoidable in the Prairies temperate latitudes. To compare the effects of feeding a leading low vicine/covicine tannin-cultivar with greater frost resistance (Fabelle) vs. the leading zero tannin faba bean cultivar (Snowbird) or soy expeller on growth performance and carcass characteristic of hogs, 504 barrows and 504 gilts were part of a commercial trial over 4 growth phases. For the entire trial, there was no difference in weight gain. Hogs fed soy expeller had greater feed disappearance but lower gain-to-feed than those fed Fabelle; hogs fed Snowbird were intermediate (Table 8). There was no difference on backfat, but hogs fed Fabelle had 1.5 mm greater loin depth than those fed soy expeller or Snowbird. Hogs fed Fabelle or Snowbird faba bean had lower diet cost, feed cost/hog, cost/kg live gain, cost/kg carcass gain, and greater income subtracting feed cost than hogs fed soy expeller.

Table 8. Overall effects of feeding a low vicine/covicine tannin-cultivar with greater frost resistance (Fabelle) vs. the leading zero tannin faba bean cultivars (Snowbird) or soy expeller on overall (d 0 to 74) daily feed disappearance (ADF), weight gain (ADG), and feed efficiency (G:F), carcass characteristic, feed cost and income subtracting feed cost of growing-finishing hogs (ISFC; Beltranena et. al., unpublished).

	Soybean expeller	Snowbird faba bean	Fabelle faba bean	P value
AFD, kg/d	2.992[a]	2.975[ab]	2.937[b]	0.020
ADG, kg/d	1.072	1.078	1.076	0.790
G:F, kg/kg	0.358	0.362	0.366	< 0.010
Carcass, kg	105.1	104.5	104.7	0.600
Backfat, mm	19.1	18.8	19.3	0.140
Loin depth, mm	62.8b	62.8b	64.3a	0.020
Diet cost, $/tonne	329.9[a]	322.3[c]	323.0[b]	< 0.010
ISFC, $/hog	81.19[b]	83.58[ab]	84.88[a]	< 0.010

5. Improvest® -Benefits That Go Beyond Improved Gain-To-Feed

Compared to castrated barrows, entire males (boars) have reduced feed intake and improved feed efficiency (Xue et al., 1997). Immunocastration allows males to perform more like boars than barrows, while avoiding boar taint in meat (Poulsen Nautrup et al., 2018). Return on investment from implementing immunocastration adds up faster as feed cost goes up. At $0.45 vs. $0.3/kg feed, the $5 vaccine cost would be paid off by 11.1 vs. 16.7 kg less feed per male pig. Claims for Improvest®, used as an immunocastration treatment, are 14 kg less feed/male, 0.3% less wean-to-finish mortality, 1% less inferior/cryptorchid pigs, 0.36 days less in barn vs. barrows (13 Canadian farm trials; Zoetis, 2019). A 0.87 kg drop in hot carcass weight and 1.7% drop in dressing percentage was countered by 2 kg heavier slaughter weight, 2 mm less backfat, 1.2% improved loin depth, 0.21 kg more ham, and 0.42 kg more shoulder for Improvest® vaccinated males vs. barrows (Poulsen Nautrup et al., 2018). Novel is Improvest® treatment of market gilts to reach heavier slaughter weight increasing carcass weight by 5.9 kg, 1% less mortality, culls, and lights, $3 lower feed cost, but 0.75 lower carcass index, 11.8 kg more feed vs. non-treated females (7 Canadian farm trials; Zoetis, 2019). Because of reduced feed intake before the 2nd dose, Improvest® males have a slightly greater requirement for amino acids and phosphorus in grower diets like untreated gilts, but this extra grower feed cost can be offset by greater intake of lower cost finisher diets fed to treated gilts (Figure 1).

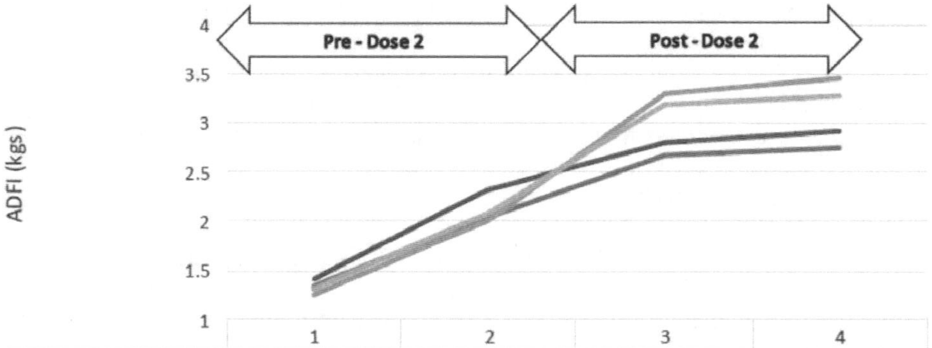

	1	2	3	4
Castrates	1.41	2.33	2.8	2.92
Gilts	1.34	2.05	2.67	2.75
Improvest Males	1.25	2	3.3	3.46
Improvest Females	1.31	2.08	3.19	3.28

Figure 1. Daily feed intakes of castrates and gilts vs. Improvest® treated males and females (Zoetis, 2019).

6. Conclusions

Our commercial scale hog trials at Lougheed, AB have shown that we can practically reduce feed energy density and thus cost and let hogs make up for the caloric difference by eating more. We have shown repeatedly that this strategy can be profitable. We have also shown that hogs with lower than recommended floor area allowance during part of the finishing phase do not benefit from feeding more energy dense rations. Hybrid rye is a good substitute for wheat grain in hog rations resulting in more pork per unit of land. Feeding a low vicine and convicine faba bean cultivar (Fabelle) resulted in slightly better growth performance and profit margin than feeding the leading low-tannin cultivar (Snowbird) or soy expeller. Benefits of implementing immunocastration go far beyond swapping vaccine cost for improved gain-to-feed.

References

Agriculture and Agri-Food Canada. 2021. Canada: Outlook for principal field crops. Aug 20, Sep 24.

Beltranena, E., and M. N. Smit. 2015. Impact of low energy diets on the bottom line. Advances in Pork Production, Vol.26, pp. 185-192.

Poulsen Nautrup, B., I. Van Vlaenderen, A. Aldaz, C. K. Mah. 2018. The effect of immunization against gonadotropin-releasing factor on growth performance, carcass characteristics and boar taint relevant to pig producers and the pork packing industry: A meta-analysis. Res. Vet. Sci. 119: 182-195. https://doi.org/10.1016/j.rvsc.2018.06.002

Smit, M. N., X. Zhou, J. L. Landero, M. G. Young, and E. Beltranena. 2021. Dietary energy level, feeder space, and group size on growth performance and carcass characteristics of growing-finishing barrows and gilts. Transl. Anim. Sci. 5. https://doi.org/10.1093/tas/txab122

Smit, M. N., X. Zhou, J. L. Landero, M. G. Young, and E. Beltranena. 2019. Increasing hybrid rye level substituting wheat grain with or without enzyme on growth performance and carcass traits of growing-finishing barrows and gilts. Transl. Anim. Sci. 3:1561–1574. https://doi.org/10.1093/tas/txz141

Smit, M. N., J. L. Landero, M. G. Young, and E. Beltranena. 2018. Effects of feeding canola meal or soy expeller at two dietary net energy levels on growth performance, dressing and carcass characteristics of barrows and gilts. Anim. Feed Sci. Technol. 235:166-176. https://doi.org/10.1016/j.anifeedsci.2017.11.013

Smit, M. N., J. L. Landero, M. G. Young, and E. Beltranena. 2017. Feeding diets with decreasing net energy value to growing-finishing barrows and gilts. Can. J. Anim. Sci. 97: 30–41. https://doi.org/10.1139/cjas-2016-0045

Xue J., Dial G.D. and Pettigrew, J.E. 1997. Performance, carcass, and meat quality advantages of boars over barrows: A literature review. Swine Health and Production 5: 21-28.

Zoetis. 2019. Improvest® overview. Dec 20.

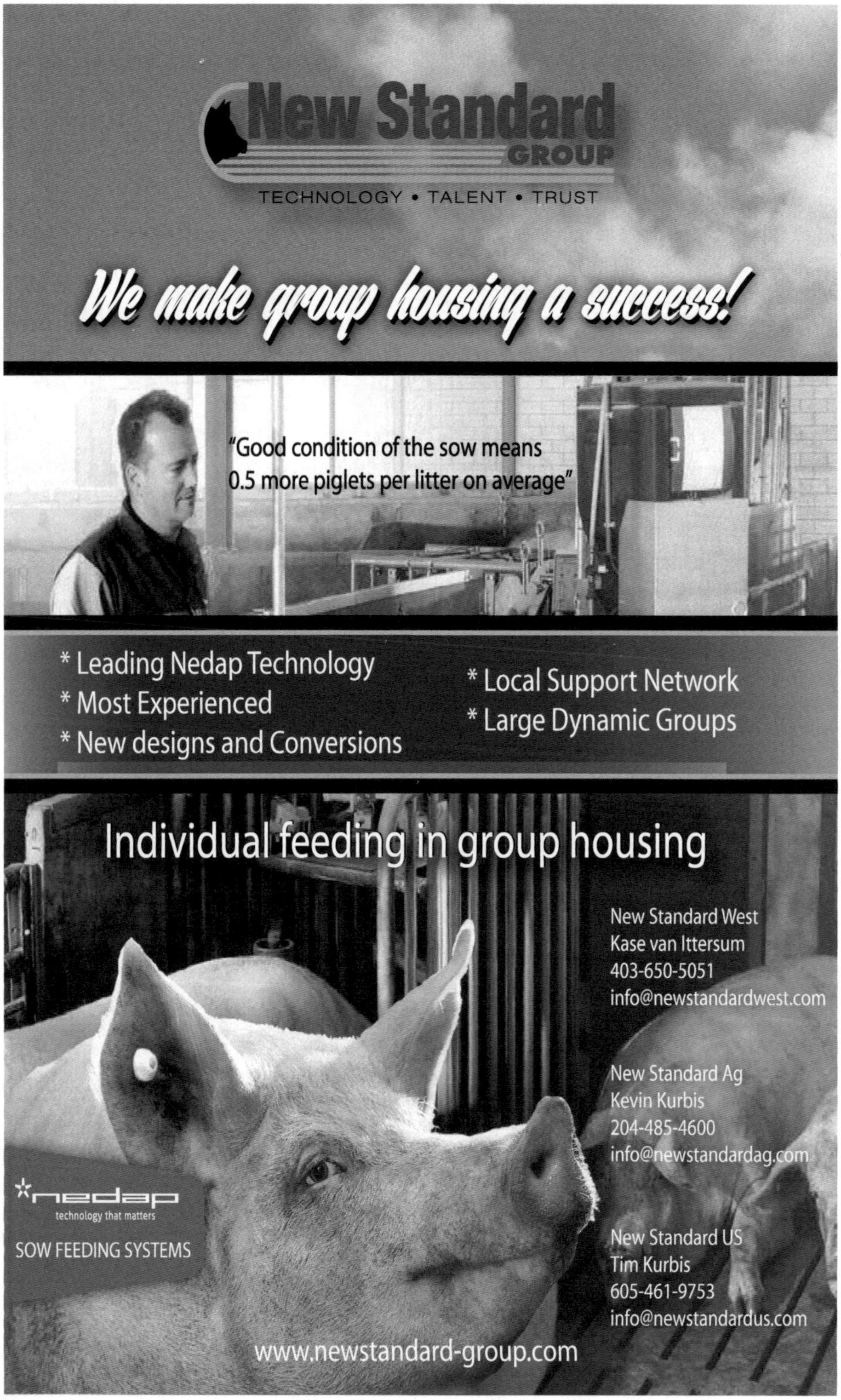
New Standard GROUP
TECHNOLOGY • TALENT • TRUST
We make group housing a success!
"Good condition of the sow means 0.5 more piglets per litter on average"
* Leading Nedap Technology
* Most Experienced
* New designs and Conversions
* Local Support Network
* Large Dynamic Groups
Individual feeding in group housing
New Standard West
Kase van Ittersum
403-650-5051
info@newstandardwest.com
New Standard Ag
Kevin Kurbis
204-485-4600
info@newstandardag.com
New Standard US
Tim Kurbis
605-461-9753
info@newstandardus.com
nedap
technology that matters
SOW FEEDING SYSTEMS
www.newstandard-group.com

Love the Ones You're With: The Secret to Employee Retention

Dennis Robles

Swine Health Professionals Ltd, Steinbach, Manitoba, Canada
drobles@shpswine.com

1. Introduction

Imagine your best employee: smart, hard-working, and very dependable when it comes to finishing the tasks, job, or projects assigned to them. A talented champion for the organization who always exceeds expectations and faces challenges with enthusiasm and a positive attitude. A team player that is going to be valued for years to come.

Now, imagine this employee being with the company for a few years. Nothing is really changing with their role and they yearn for more growth, new skills, and challenges. As anyone else would, they look for recognition and appreciation for their work, even though they would not seek it or ask for it—they just wait. They may drop some hints in passing or be direct at times, but the manager does not have the skill set to acknowledge these hints or requests. Now the employee feels misunderstood, overlooked, and starts to think and look outside of the company for other opportunities. The team would get bonuses occasionally, and the company holiday parties are okay, but the employee asks, what else is out there?

Most of the organizations in the swine industry are in this predicament. Shortage in labour is the biggest challenge for pork producers in Canada and the US.

> *"Canadian livestock production and packing is facing a mounting labor shortage. This shortage has been brought on by many factors, including an aging work force, a dramatic reduction in local youth enrolling in agriculture-related programs/farm careers, and an inability to compete in the labor market with other sectors. This new labor environment is directly impacting hog production. – National Hog Farmer Oct.29,2021"*

That inability to compete in other labour markets and keep employees is the number one focus that organizations must have. Gallup is a global analytics and advice company that helps organizations be their best. Gallup's research shows that 48% of US employees are actively seeking other job opportunities (Gallup, The Great Resignation is the Great Discontent, July 22, 2021). Companies assume that this is inevitable, but results show that 52% of voluntarily exiting employees say their manager or organization could have done something to prevent them from leaving their job. One might argue that this is data from across the border, however, we cannot deny that this situation is the same in Canada.

What are the factors that make the talented employees stay, engage, and succeed with the organization? It depends on the manager or the leader. The best leaders don't wait for situations where the employee is dissatisfied to the point of leaving.

They find preventive measures and strengths-based management approaches as the best way to keep employees.

2. What Employees Want

"People want purpose and meaning from their work. They want to be known for what they're good at" (Gallup, Employee Engagement and How to Improve il, 2021).

Employees, especially talented employees need more than a fleeting warm-fuzzy feeling and a good paycheck (even if it helps them respond positively to surveys) to invest in their work and be committed to the company. They want to know what makes them unique and they want to feel appreciated. They want to have a good relationship with their immediate manager/supervisor, who can support and coach them to develop. One of Gallup's biggest findings in its vast research shows that 70% of a team's engagement variance is attributed to the team leader or manager.

The changing demands of the workforce are shown by J Clifton & J Harter in the book *It's the Manager*:

THE PAST	OUR FUTURE
My Paycheck	My Purpose
My Satisfaction	My Development
My Boss	My Coach
My Annual Review	My Ongoing Conversations
My Weaknesses	My Strengths
My Job	My Life

Figure1. Changing demands of the workplace.

To keep talented employees, we need people management skills to engage workers. Gallup defines engaged employees as those who are involved in, enthusiastic about, and committed to their work and workplace.

In the swine industry, particularly at the barn level, the same principles apply. To keep employees, we must keep them engaged and committed in order to achieve more for the company. Clear, consistent, and constant conversations about expectations and performance achievements drives high engagement. Career development programs and team or personal growth strategies within the company work to elevate engagement. We need to have team leaders and managers that are equipped with the "right tools" to be great at coaching their teams instead of being "the boss". Trust and respect must be developed in all levels of the organization. Larger companies can start with strong team-building efforts with all the team members in their respective departments. Knowing what each person contributes to the team in terms of their talents and skills fosters good relationships, trust, and respect.

3. Where is the Disconnect/Gap?

Unfortunately, most employees are not engaged or are actively disengaged according to Gallup's state of the GlobalWorkplace:2021 Report. A global employee engagement rate of 20-34% in the US and Canada is very costly:

- The lost productivity of not engaged (those not who are not are involved, enthusiastic, and committed) and actively disengaged (INSERT DEFINITION HERE) employees is equal to18% of the annual salary.
- For a company of 10 employees with an average annual salary of $40,000, disengagement costs $72,000 a year.
- Replacing workers can cost ½ to 2x the annual salary. So, in the previous example, it costs $7,200 to keep an employee that is disengaged and $20,000 to $80,000 to replace them.

Great managers must have the skillset to fill the 12 elements of engagement ranging from knowing what is expected of them to having the opportunity to grow, learn, and develop. This will make the team productive with team members that are more resilient and have high well-being. This is not a difficult task, but we must equip our managers to be able to do this right. Managers need to know how to conduct meaningful conversations, set expectations, and create accountability within their team.

It's not all about money. Though pay is important for employees' decisions to stay or leave, it is not the only solution to keep talent. There are well-paid employees that leave due to disengagement in their jobs. Gallup data shows that it takes more than a 20% pay raise to lure employees away from a great manager who engages them, but it costs almost nothing to poach disengaged workers. Disengagement rather than pay is a better predictor of employees' level of loyalty or retention. Perceptions of the fairness of pay incentives vary between employer and employee. Those employees who feel that their pay and incentives are fair are more motivated and contribute more to the company. However, results show that only 22% strongly agree with those feeling or perceptions (Gallup, How to Align Your Employee Compensation and Talent Strategies, 2021).

4. The Role of The Manager/Owner on Bridging the Gap

Having that much impact, 70% on employees' engagement (retention), the manager's or team leader's role is huge. Finding & hiring who is going to be the manager or team leader is one of the biggest challenges a company has. Developing managers and team leaders is one of the most important efforts that great companies focus on. As they find and develop managers, they emphasize on the following.

I. Creating a culture of engagement

When employees are engaged, they are set up for success. They are motivated, they know what is expected of them, and they strive to contribute more. A great manager provides employees with what they need to do their work well, helps employees manage their workload, builds great collaboration between teammates, and makes meaningful working relationships of trust and respect.

II. Developing a high well-being team

Gone are the days of the work-life balance. We are now in the age of work-life integration where we bring our best self to work (because our home is great) and our best self to home (because we have great work). Authentic relationships between managers and their team helps the team work better. Whatever life throws at someone within the team can be managed and adjustments can be made to overcome those challenges.

III. Establishing a culture that celebrates each person's talents & strengths

Self-awareness is the start of personal development. With Gallup's CliftonStrengths assessment, anyone can be aware of the talents and strengths they bring to the world. Having a list of talent themes and a report that individually describes those talents is a very powerful tool a person can use to live a fulfilled life. And, sharing those talents with others—family, friends, and co-workers—and using them every day can result in an excellent quality of life. The CliftonStrengths assessment report is a tool managers can use to lead, help, and coach their team.

People who learn to use their strengths every day at work have 7.8% greater productivity and teams who receive strengths feedback have 8.9% greater profitability.

5. Conclusion

The struggles in labour shortage are a real challenge to the whole industry—pork producers, packers, service providers and stakeholders. We must look for new strategies to attract and keep talented employees for continued growth and productivity. In times of challenges, whether it is disease- or production- or market-driven, people are the cornerstone of our organizations. These are the times when they need the most help and support from well-trained and great managers and team leaders. Getting them engaged, making sure that they have high well-being, and appreciating their talents are basic requirements for the current workforce. It costs to replace an employee and returns are great if you have a plan to keep the ones you have.

When Stress is More Than a Season

Lesley Kelly

The Do More Agriculture Foundation

Farming can be both a rewarding and demanding occupation where farmers experience highs and lows and mounting stress. Everyone feels stress and there can be lots of negative stress on the farm, including ongoing unhealthy family or employee dynamics, debt loads, long hours, and unexpected interruptions. But there is also good stress that fuels and motivates us.

What is Stress?

According to the Canadian Mental Health Association, stress is a normal response to situational pressures or demands and is part of everyday life. Often referred to as "fight or flight," stress causes our bodies to release hormones which make our blood pressure, heart rate and blood sugar levels rise. A nearly constant state of stress, or chronic stress can lead to mental health problems and medical issues.

Is All Stress Harmful and Negative?

A small amount of stress can be good for us because it gives us that boost of adrenaline that motivates us to accomplish tasks. We see good stress on the farm like the last push of getting the crop planted or seeded before a rain comes or helping a heifer birth a healthy calf.

When stress becomes overwhelming and prolonged without the opportunity to recover, it can be harmful and risks for mental health problems and medical problems increase. Long term stress increases the risk of mental health problems such as anxiety, depression, substance abuse problems, sleep problems. Stress also increases the risk of medical problems like muscle tension, chronic headaches, weakened immune systems, high blood pressure, cardiovascular disease and stroke.

Recognizing Signs and Symptoms of Stress

Physical

- Headaches
- Chest pain
- Beating heart
- Grinding/clenching teeth
- Fatigue

Emotional

- Difficulty controlling emotions
- Increased irritability or angry blow-ups
- Frustration
- Depressed
- Impatient
- Low self esteem
- Suicidal thoughts
- Change in diet and sleeping habits

- Substance use increase
- Lack of concentration
- Forgetfulness
- Procrastination
- Withdrawing from others and activities
- Trouble adapting to changing situations

Because stress is a normal part of life, everyone experiences it. However, the frequency, duration and intensity will be different for each person and therefore, how each person manages their stress will be different too. Here are some quick tips that you can do in your office or cab of the tractor to help manage stress and help you weather the storm of stress:

- Stay Focused on What You Can Control: When it seems as though things are outside of our control, we can focus on what we can control and that can help us feel empowered. For instance, you may not be able to control how much rain we get this summer or who contracts COVID-19, but you can control things like hand washing and disinfecting machinery and shop areas where people come into contact with. From a business perspective, you can control keeping up to date with record keeping, communicating with your team and family, and self-care activities.
- Stay Connected: Check in with your family and friends and plan on how you can stay connected through this time, whether that's through digital options, phone calls, or social distancing. Have people, whether it is your

spouse, a family member or a friend, you can talk and express your feelings to.

- Talk it out with a friend: Having supportive people in your life is the key to stress management. This may mean reaching out to your existing network or expanding your network by joining an organization, attending a support group or obtaining professional help.
- Talk yourself through it: Sometimes calling a friend is not an option. If this is the case, talking calmly and positively to yourself can be the next best thing. Tell yourself why you are stressed out, what you have to do to complete the task at hand, and most importantly, that everything will be okay.
- Tune Out: 24-hour news available at our fingertips can be too much at times. It's important to stay informed but feel free to take a media break from updates if they start to negatively impact your mental health.
- Maintain Your Health: Eating right, staying active, getting fresh air and keeping hydrated are a few ways to help maintain your physical health, which will also have positive impacts on your overall mental health.
- Do Some Good: Helping others like volunteering your time and energy can help you feel good as well. Plus, it helps us weather the storm and get through hard times as a community and help others and yourself not feel alone.
- Embrace the Good: We learn resiliency by going through adversity. In hard times, there are always positive things that shine through about ourselves, our neighbours and community. These things may be hard to

see but they are there. When you find them, embrace these opportunities to see, grow and learn from.

- Don't Forget to Play: Take this time to do something that makes you feel good, like trying a new recipe, listening to music, or getting outside and playing a sport with your family.
- Know it is Okay to Feel Many Different Things: There is no 'right' way to handle unprecedented times like we are in. Anxiety, fear, worry, uncertainty, and grief are all normal feelings to have. It's important to be kind to yourself and seek help if these feelings start to take a toll and cause disruptions to your daily life.

Prioritize and Adapt Your Self-Care Routines

Self-care looks and feels different for everyone and changes with the seasons. Simply put, self-care are activities that help your physical, mental and emotional health and help you take care of yourself, which will then help you take care of others around you. Some of your self-care routines may be disrupted right now like going to the gym or hanging out at your favourite restaurant with your friends. Consider what you can do differently and take this chance as an opportunity to try something new.

In the moment

- Talk yourself through that stress or with someone else
- Take a moment: Get off the tractor, put down the phone or radio, go for a walk, move

- Breathing exercises: Just focusing on your breath or changing the way you breathe can make a big difference to your overall stress level. Breathing techniques can calm your body and your brain in just a few minutes. While there are many breathing exercises, a few simple ones include:
    - Breathe in through your nose and watch your belly fill with air. Count slowly to three as you inhale. Hold for one second and then slowly breathe out through your nose as you count to three again.
    - Breathe in through your nose and imagine inhaling peaceful, calm air. Imagine that air spreading through your body. As you exhale, imagine you're breathing out stress and tension.
- Write things down such as why you are stressed, a priority list, goals, etc.

Long term habits

- Sleep, hydration, diet, and exercise
- Getting help with jobs/activities that are stressful
- Taking off farm breaks
- Professional support

If stress begins to it impact your mental health and your ability to carry out day to day activities and tasks, it may be time to seek professional help. Visiting your family doctor or the resources below is a great place to start.

Resources

www.domore.ag : The national voice and leader for mental health in agriculture that contains information, and resources across Canada, as well as local, regional and provincial services and resources.

Mental Health Line: 211

Manitoba	**Saskatchewan**	**Alberta**
Manitoba Farm, Rural & Northern Support Services: 1-866-367-3276 Supportline.ca	Farm Stress Line: 1-800-667-4442 Mobile Crisis Services: 1-306-757-0127	Mental Health Hotline: 1-877-303-2642 Mymentalhealth.ca

Improves piglet and sow health
Jefo's specific blend of protected organic acid for piglet and sow, will help improve gut health and reproductive performance.
Increases growth rate and lowers incidence of diarrhea at weaning
Improves litter size and body condition
Reduces vulval discharge and UTI
Jefo
For more information, contact your specialist :
Rhod Ross
North Alberta / Saskatchewan
(403) 702 6709 | rross@jefo.ca
Ron Jones
Alberta / Saskatchewan
(403) 804 1205 | rjones@jefo.ca

Benchmarking Production Trends & Mortality

Greg Bilbrey, MS

PIC, Hendersonville, TN, USA

1. Introduction

Benchmarking individual company or producer key swine production metrics against industry production trends can be an effective way to evaluate and improve performance, cost efficiency and ultimately, profitability. Agri Stats, MetaFarms, PigCHAMP, and Swine Management Services (SMS) are some of the professional benchmarking options currently available in the industry. Lenders, genetic companies and feed or nutrition suppliers may also offer benchmarking services or information. Calculation methods for key metrics, diet energy levels, pig start and ending weight differences, period or season and other factors must be considered and may influence comparisons and benchmarking results.

In this presentation and article, we will examine current production trends and differences using data from PIC, Agri Stats and other industry benchmarking sources. Emphasis will be placed on sow and finishing mortality trends from the data sources used.

2. Sow Production

There is a wide range in sow farm productivity within the North American swine industry. The most recent Agri Stats quarterly data has an average of 27.45 Pigs

Weaned per Mated Sow per Year (PSY) with a range of 23.2 to 32.0. PIC customer data shows certain farms at 23 PSY while other farms average 35 PSY. Differences may be attributed to prevalence or lack of health challenges (PRRS, PEDV, etc.), health cleanup strategies, labor availability, management capability, etc. Table 1 shows actual sow farm performance data from multiple production systems. PSY in Table 1 ranges from 25.80 to 33.00. Advantages or disadvantages in the other key metrics help explain the differences in PSY. Benchmarking multiple key metrics across companies allows us to compare and evaluate metrics independently to look for opportunities for improvement.

Table 1. Sow performance metrics comparison across companies.

Period: July-August-Sept 2021

	Production System				
	Sys1[a]	Sys2[a]	Sys3[b]	Sys4[c]	Sys5[c]
Pigs/Mated Sow/Year	33.0	25.8	28.4	29.3	31.7
Litters/Mated Sow/Year	2.51	2.32	2.37	2.46	2.45
Farrowing Rate, %	91.9	86.1	87.7	93.1	87.3
Total Born/litter	16.0	14.6	16.0	16.2	15.3
Born Live/litter	14.7	13.4	14.1	14.3	14.0
Pre-wean mortality, %	10.5	16.2	14.9	16.3	8.1
Weaned/Sow Weaned	12.9	11.0	12.0	11.9	12.9

[a] MetaFarms data; [b] PigCHAMP data; [c] PigKnows data

Table 2 shows sow farm data within a single system. Data in Table 2 shows us there are opportunities and benefits to be gained from benchmarking within a production system. Again, we would compare and evaluate individual metrics to look for opportunities for improvement.

Table 2. Sow performance metrics comparison within system
Period: July-August-Sept 2021; MetaFarms Sow Farm Comparison Report

	Production System				
	Farm 1	Farm 2	Farm 3	Farm 4	Farm 5
Pigs/Mated Sow/Year	34.5	35.0	33.1	31.0	30.2
Litters/Mated Sow/Year	2.56	2.50	2.53	2.44	2.50
Farrowing Rate, %	94.0	91.5	92.2	87.6	95.3
Total Born/litter	16.4	16.5	15.2	15.5	15.9
Born Live/litter	15.1	15.2	14.1	14.3	14.2
Pre-wean mortality, %	10.8	9.4	9.1	13.4	11.5
Weaned/Sow Weaned	13.2	13.5	12.9	12.4	11.9

Table 3 shows performance data from PIC's quarterly production summary comparison. The PIC comparison shows numerical and percentile rankings of selected systems for key metrics. In this example, we can compare systems by metric, or we can also look for the main opportunities by system. For example, System 1's lowest ranking is in Sow Mortality. Actions could be taken to determine reasons for sow loss and then make the improvements necessary. System 2 ranks the lowest in PSY out of the four systems listed. Since PSY is a composite of other metrics, we would evaluate those to look for ways to improve PSY. Pre-wean mortality shows as the next lowest ranking for System 2 so that tells us where we might first pursue opportunities.

Table 3. PIC quarterly production summary comparison.
Period: 2nd Quarter, April-May-June, 2021; 46 systems; ~1m sows

	Top 25%	System1	System2	System3	System4
Pigs/Mated Sow/Year	32.7	33.1	27.0	29.7	27.7
Ranking/46		2	31	13	24
Percentile		96%	33%	72%	48%
Farrowing Rate, %	91.9	91.9	88.1	86.7	88.8
Ranking/46		5	22	29	17
Percentile		89%	52%	37%	63%
Total Born/litter	15.8	16.0	15.7	16.1	14.6
Ranking/46		8	17	6	44
Percentile		83%	63%	87%	4%
Pre-wean mortality, %	9.7	9.4	15.2	13.2	14.6
Ranking/46		4	25	14	22
Percentile		91%	46%	70%	52%
Sow Mortality, %	10.5	11.8	11.7	14.2	15.6
Ranking/46		24	23	31	35
Percentile		48%	50%	33%	24%

Table 4 shows trends of the Top 25% in PSY in the PIC quarterly comparison. The same information is shown in graph form in Figure 1. This data shows how the top 25% in PSY trends over the last eight quarters for five key performance metrics. We see the top 25% in PSY improving over time in PSY, Total Born, Farrowing Rate and Pre-wean Mortality. Annualized Sow Mortality increased quadratically with the highest mortality rate observed in Q4 2020. This was also the highest quarter for Pre-wean Mortality.

Table 4. PIC quarterly production summary top 25% trends.

Period: 2nd Quarter, April-May-June, 2021; 46 systems; ~1m sows

	PMSY	Tot Born	Far Rate %	PWM%	Sow Mort %
Q3 2019	30.7	15.6	89.8	10.2	8.3
Q4 2019	31.2	15.2	88.7	10.0	11.2
Q1 2020	31.0	15.7	88.8	10.2	10.9
Q2 2020	30.9	15.8	90.1	10.5	11.3
Q3 2020	31.2	15.8	89.7	11.2	11.3
Q4 2020	30.5	15.5	88.6	12.1	14.6
Q1 2021	30.9	15.6	91.1	10.0	10.7
Q2 2021	32.7	15.8	91.9	9.7	10.5

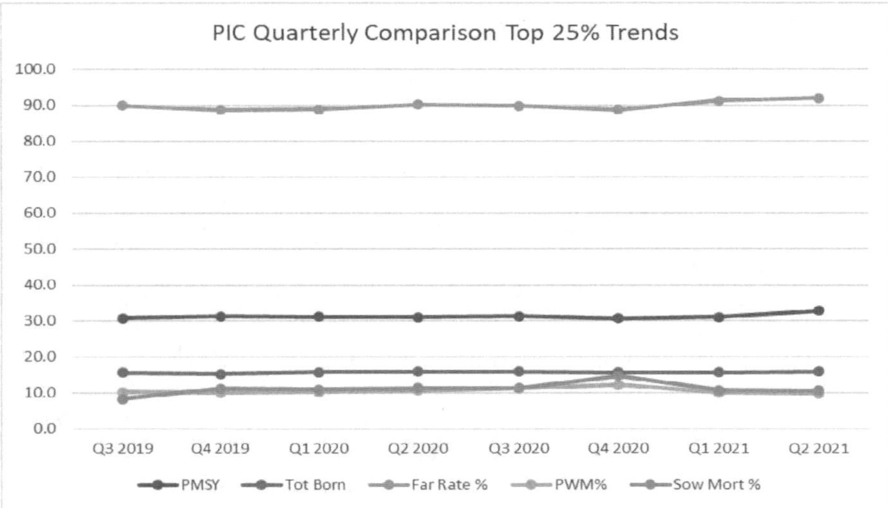

Figure 1. PIC quarterly comparison top 25% trends

2.1 Sow Mortality

The industry increase in sow mortality is currently being discussed in many publications and tech letters. Agri Stats data shows weighted average annualized sow mortality increased from 11.1% in Q1 2015 to 14.4% in Q4 2020. PIC data in Table 4 shows an increase of 2.2 points over the most recent eight quarters for the top 25% in PSY. Reasons for the increases may include the industry shift to more pen gestation, less antibiotic use, enhanced welfare standards and increased euthanasia, enhanced cull sow requirements, an increase in pelvic organ prolapses and more extreme PRRS challenges in 2020-2021.

3. Grow-Finish Production

As with sow production, multiple options exist for benchmarking or comparing grow-finish production data. Most producers keep group closeout data via their own Excel spreadsheets or commercially available records systems such as MetaFarms. We again need to be sure we are making "apples to apples" comparisons as factors such as calculation methods, start and end weights and diet energy levels can greatly affect performance levels and information. We also need to remember that best performance does not equate to best cost or profit and our goal is not necessarily to rank the best in a specific metric.

PIC provides a quarterly grow-finish benchmarking comparison for customers. Raw closeout data is collected and audited to remove outliers or irregular data. All metrics are calculated using raw data in the same calculations for all participants. Weights are adjusted as follows: nursery 12-50 lb (5.44 – 22.68kg); finishing 50-280 lb (22.68 – 127kg); wean-finish 12-280 lb (5.44 – 127kg). Table 5 shows a sample participant performance table from a Q2 2021 report.

Table 5. PIC Q2 2021 WF benchmarking comparison, cumulative performance spread

	Closeouts #	Closeouts %	Number Pigs Sold	ADG lb/d	Feed: Gain FCR	% Mortality	PIC Opportunity Cost, $/hd	PIC Opportunity Cost vs. ALL OTHERS
SAMPLE	**39**		**121,691**	**1.65**	**2.37**	**9.00%**	**$8.78**	**$0.88**
Top 1%	0	0.00%	0	0.00	0.00	0.00%	$0.00	$0.00
Top 5%	0	0.00%	0	0.00	0.00	0.00%	$0.00	$0.00
Top 10%	0	0.00%	0	0.00	0.00	0.00%	$0.00	$0.00
Top 25%	5	12.82%	19,750	1.82	2.30	4.60%	-$0.72	$1.51
Top 50%	12	30.77%	41,274	1.76	2.32	5.10%	$1.39	$0.67
Remaining 50%	27	69.23%	80,417	1.59	2.40	11.00%	$12.58	-$1.14
ALL OTHERS	**353**		**649,145**	**1.72**	**2.50**	**5.50%**	**$7.90**	
Top 1%	4	1.13%	4,904	1.88	2.18	2.80%	-$7.56	
Top 5%	20	5.67%	25,246	1.88	2.23	2.70%	-$5.75	
Top 10%	40	11.33%	47,908	1.87	2.26	3.00%	-$4.54	
Top 25%	97	27.48%	126,863	1.84	2.30	3.40%	-$2.24	
Top 50%	191	54.11%	290,447	1.81	2.36	3.90%	$0.72	
Remaining 50%	162	45.89%	258,698	1.65	2.61	6.80%	$13.72	
TOTAL	**392**		**770,836**	**1.71**	**2.48**	**6.00%**	**$8.04**	

In this example we see an actual participant nicknamed "SAMPLE" compared to ALL OTHERS without SAMPLE included. SAMPLE had 39 closeouts for Q2 2021 compared to 353 for ALL OTHERS. PIC calculates an opportunity cost for all closeouts. Average Daily Gain (ADG), Feed Conversion Ratio (FCR) and Mortality are compared to the top 25% data from the previous year. If ADG for a closeout is below the top 25% ADG from the previous year, there is a cost of extra facility days applied ($0.18/day in Q2 2021). Likewise, if FCR is worse than the top 25% FCR from the previous year, there is a cost applied for the extra volume of feed required ($195.00/ton Q2 2021). Mortality cost is simply the average of placement and finishing costs per head. Closeouts are ranked according to total opportunity cost values and distributed into the percentile categories shown. We first compare SAMPLE in bold to ALL OTHERS in bold: 1.65 ADG vs. 1.72, 2.37 FCR vs. 2.50, and Mortality 9.00% vs. 5.50%, respectively. We also see 69.23% of SAMPLE's closeouts fall into the remaining 50% distribution while 30.77% are in the top 50%

category. Obviously, it would be better if SAMPLE could move more of their production upward into the top percentile groupings. We also notice SAMPLE had no closeouts fall into the top 10%, top 5% or top 1% categories and only five closeouts or 12.82% of their production made it into the top 25%. This type of analysis allows us to evaluate variation and opportunities within a system instead of just looking at weighted averages of all production. Looking further at the data, we see SAMPLE's Remaining 50% ADG and Mortality are not as good as ALL OTHERS': ADG 1.59 vs.1.65 and Mortality 11.00% vs. 6.80%. Neither SAMPLE nor ALL OTHERS performed as well as our target top 25% from the previous year. Therefore, SAMPLE had an opportunity cost of $8.78/head while ALL OTHERS had an opportunity cost of $7.90/head. The last column is simply the difference between SAMPLE and ALL OTHERS opportunity costs/head. It is important to note that these opportunity costs are based on performance only and do not represent SAMPLE's or any participant's actual costs. Therefore, the opportunity cost values are theoretical and are used for evaluation and discussion purposes. Participants in the PIC comparison also receive ranking and trendline graphs for each performance metric. Examples are shown in Figure 2.

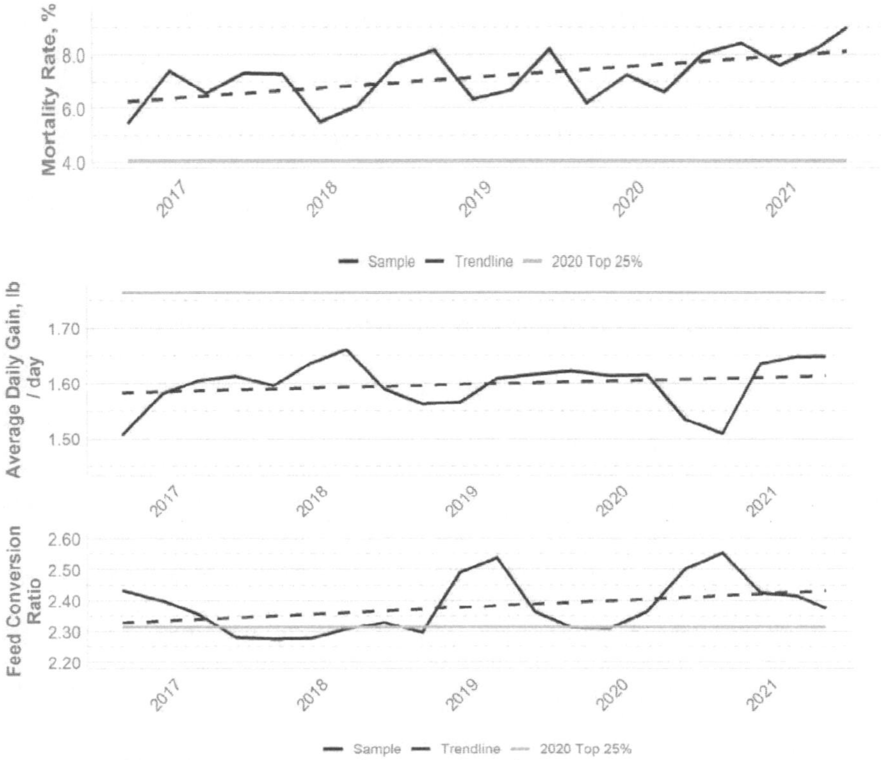

Figure 2. PIC sample quarterly performance trends

Table 6 shows Q2 2021 PIC and industry nursery, finishing and wean-finish production data as well as a percent change for PIC data vs. the previous three years.

Table 6. Industry and PIC Q2 2021 GF benchmarking summary & trends

	Top 10%	Top 50%	Remaining 50%	Average	% Change vs. Previous 3 yrs	Other Industry Benchmarks #1	#2
NURSERY							
ADG, lb/d	0.95	0.88	0.77	0.82	98%	0.82	0.85
FCR	1.26	1.42	1.63	1.53	101%	1.54	1.52
Mortality, %	1.60	2.00	8.60	5.30	196%	6.00	4.40
FINISHING							
ADG, lb/d	2.28	2.13	1.95	2.05	100%	1.88	1.96
FCR	2.44	2.56	2.76	2.64	102%	2.77	2.82
Mortality, %	2.50	3.30	5.50	4.20	140%	5.80	5.20
WEAN-FINISH							
ADG, lb/d	1.86	1.80	1.65	1.71	102%	1.68	1.69
FCR	2.26	2.36	2.59	2.50	100%	2.69	2.57
Mortality, %	3.00	4.00	7.30	6.00	101%	11.00	6.70

3.1 Grow-Finish Mortality

Table 6 shows significant mortality increases in PIC nursery and finishing closeout data compared to the previous three years. Obviously, health challenges in placed pigs as well as disease breaks in nurseries and grow-finish barns can result in large fluctuations in mortality. Mortality has a huge impact on overall system cost and profitability. With permission and assistance from Agri Stats (Luke Gorski, Agri Stats, Inc., Fort Wayne, IN, USA), a profit factor analysis was done to determine the advantages of the top 25% in farm profit. Data is shown in Table 7. This analysis involved calculating the variance of the top 25% in profit to the average for fourteen key metrics from 2008 through 2019. The fourteen key metrics were ranked 1 – 14 and split into a top seven and bottom seven. Finishing mortality was one of the top seven factors in nine of the twelve years evaluated.

Table 7. Profit factor analysis – Agri stats data, 2008-2019, with permission.

Advantages of the Top 25% in Profit
Top 7 Advantage Rankings by Year and All Rankings Summarized 2008-2019

PROFIT RANK	SALES $/CWT	% CULL LBS	FIN WT	FIN AGE	FIN MORT%	FIN ADG	FEED CONV	CAL FC 50-270	FINISH $/TON	COST $/CWT	$/WEAN PIG	LMSY	BORN LIVE	PRE-WN MORT%
2019	7	1			3		6	4		5	2			
2018		1			5				4	6	2	7		3
2017		1			6				4	5	2	7		3
2016	7	1							4	5	2	6		3
2015		2			1				4	5	6	7		3
2014		2			3				6	5	4	7		1
2013		1			2		7			6	4	5		3
2012	7	1							6	4	3	5		2
2011		1			3				6	5	4	7		2
2010		1			2	7			6	5	3			4
2009	3	1			2	6			7	5				4
2008	4	1	7			6			3	5				2
2008 -			10	11		10	12	10				8	11	
2019	7	1			5				6	5	5			4

Top 25% in Profit variances to average ranked 1 - 14 on largest % deviation to average.
Indicates advantages of the top 25% in profit to the population by year.

4. Conclusion

Benchmarking can be used to improve efficiency, cost and profitability of swine production operations. Multiple options exist for benchmarking both sow farm and grow-finish performance. Sow performance is increasing though there remains a lot of variation in performance and opportunity across the industry. This is true also for nursery and grow-finish performance.

Always open to new business opportunities. On an expansive global payments network.
With our extensive global network, we can help the Pork industry to manage and simplify international payments.
Contact us:
Matthew Kurlapski
Area Sales Manager - Canada
(514) 516-3243
Matthew.Kurlapski@wu.com
WesternUnion\\WU
Business Solutions

HIPRA
ARE YOU READY?
WE ARE
Piglet ID vaccination with MHYOSPHERE® PCV ID has never been so precise, so easy or so quick
If you are curious contact your veterinarian and/or Jonathan McKnight, our territory manager:
jonathan.mcknight@hipra.com
www.mhyopcv2.com
www.hipradermic.com

Are Your Piglets Ready to Fly the Coup? Setting Your Pigs up for a Successful Weaning, On and Off the Sow Farm

Jessica Law

Prairie Livestock Veterinarians, Red Deer, Alberta, Canada.
Email: Jessica@prairieswinehealth.com

1. Introduction

Weaning is one of the most crucial periods in a piglet's life. This is when all of that hard work in the sow farm and the preparation at the nursery come into play. It is likely the most stressful period for both sows and piglets, depending on the practices and the logistics at the farms. Accomplishing a successful weaning becomes a day-to-day practice and overtime little things get missed. Cutting corners in this event can lead to big losses downstream. Setting your team and your piglets up for success is crucial for pig welfare and the bottom line.

2. From Day One

Weaning pigs is affected by the work done even prior to the sows farrowing. There are numerous factors that impact a piglet arriving as a full value pig to market, and a critical portion of this is the immune system. Lactogenesis, ability of the mammary gland to create milk and its components, starts at about day 105 of gestation, within this period antibodies begin to accumulate within the tissue and are synthesized locally. Ensuring the sow can perform this is critical, this means providing high quality feed and sufficient and clean water access (2L/minute for up to 25L/ day). The colostrum produced in these mammary glands will provide immunity to the piglet for the first 10 days of life and protection into the nursery period. Once the colostrum is produced and the piglets are born, the piglet has 12 hours to acquire the much-needed colostrum for energy, nutrition as well as immunity to ensure their survival. Piglet weight at birth, vitality and colostrum consumption all contribute to their survival and growth through to finishing. Within the first 24 hours piglets require 160g of colostrum per kg of birthweight for survival. Ensuring each piglet acquires this through teat training, split suckling and colostrum collection has an incredible impact on piglet survivability.

Pre-farrowing vaccinations should be reviewed with your veterinarian to ensure that the appropriate vaccines, timing, and administration are occurring and are both advantageous to the immune protection of the piglet as well as minimize stress on the sow during gestation.

Once the piglets are on the ground, dried off and suckling, the next 3-4 weeks involves great attention to their health and management of that litter. Ideally all disruptions such as fostering, processing and vaccinations are done early on to minimize stress. Fostering should be done within the first 24 hours, but after the piglets have suckled from their own sow. No piglet should move without first having colostrum from their own mother. Piglets should only be moved on or off the sow, not both and after 24 hours, moving piglets should be halted unless runt litters or nurse sow litters are necessary. Aim to cascade foster (move piglets in a stepwise manner) if a nurse sow is needed, rather than moving 1-day old piglets onto a sow that has just weaned her litter as her milk production will suffer as a result of the piglet's lack of energy and ability to draw down her teats as much as a 12 or 15 day old litter might be able to.

All of these initial inputs and processes can impact the health at weaning and decrease the amount of work needed to hold-back or treat piglets later on. If piglets are carefully monitored and smalls litters are made up early on (day 7-10) rather than the day of weaning, fewer litters and pigs will be disrupted and weaning weights and growth on the sow will improve significantly.

3. Transport

Ensuring transport is optimized can make a significant difference on dead on arrival numbers as well as time it will take the pigs to get started on feed. Guidelines for weaned pig transport can be difficult to interpret as the research is limited in North America for our extreme temperatures. It is clear the pig movement. The literature at this time cites 15°C as "cold/cooler" temperatures, which most producers know that this is far from our reality. In the event that producers do have control of logistics, loading in the warmer times of day in cooler months and ensuring plenty of bedding is used to avoid the piglets from contacting the cold trailer is essential. In the event that weather warnings are advising for humans to stay inside, it is ideal for loads to be rescheduled, that being said, it is highly unlikely in most cases that space availability allows for this. Focusing on the aspects that can be controlled in transport is critical to ensure piglets experience as little stress as possible.

4. Placement

On arrival to nursery rooms, ensure that rooms are wash, disinfected and dried in advance. Allowing for at least a day or two dependent on the month is important to ensure the rooms are thoroughly dried. Drying is essential for pathogen reduction and inactivation as well as appropriate disinfection use and application. Turning the heat on at the time of placement is inadequate and the rooms must be up to temperature to ensure concrete flooring and metal fixtures are also warmed up at this point. Drinkers that have been sitting in a room at 15°C or colder will be too cold for use and the piglets will avoid drinking. Allowing for the room to warm up prior to placement well help minimize unexpected consequences. Placing mats for floor feedings is helpful, especially in the event piglets are not fed creep feeds prior to weaning.

Some farms will vaccinate piglets at the time of arrival to the nursery or perhaps in the farrowing crate. Others will vaccinate a few days after arrival. Whichever protocol is used by your site, it is important to work with staff to ensure that low stress handling techniques are used. Needless injectors can be used to minimize the tissue damage and inflammation inflicted by vaccination. Needle hygiene and syringe hygiene should be reviewed regularly to insure that disease is not spread or induced unintentionally by these interventions (iatrogenic spread).

5. Post-weaning Care

Additional heat lamp and creep areas can be used in the winter months if pigs experience diarrhea or are smaller than average. Other forms of supportive care include use of electrolytes in the water to provide additional nutrients until the pigs start on feed. We routinely use sodium salicylate to prevent fever, illness and minimize inflammation in piglets for five days post weaning.

Working with your veterinarian to ensure your staff are well aware of the signs of illness that may indicate mass treatment is helpful. Having upper limits of intervention in regard to number of pigs individually treated for clinical disease such as diarrhea, respiratory disease and lameness may help staff notify the veterinary team in advance of disease progressing too far. Early treatment is key to managing disease especially post-weaning as these illnesses will cause tissue damage that may affect the piglets long past the nursery stage.

6. Conclusion

There are numerous intervention points from the time of birth until a few weeks after placement in which the weaning period can be optimized. Focusing on piglet quality and piglet care with an individual and herd lens will help your team ensure that the stress of weaning is minimized. Through understanding the stressors experienced by piglets and the different points in time in which disease can be influenced or worsened, the weaning period can become less stressful and more routine for you and your pigs.

Impact of Wean Age on Reproduction and Grow Finish Performance

Brady McNeil, Amanda Cross, Jason Schneider, Tom Rathje and John Sonderman

DNA Genetics, 4438 Old Mill Court, Columbus, NE 68601, USA
bmcneil@dnaswinegenetics.com

1. Introduction

Wean age over the last 30 years has changed dramatically within different production systems. The slow but increasing change in wean age has eliminated the need for early wean strategies on most farms. Wean age increased from 18.5 days in 2005 to 20.7 days in 2020 (PigCHAMP Benchmarking USA year summary, 2020 and 2005). As average wean age has increased, wean age range has also increased from 16.7 - 20.5 days in 2005 to 18.7 - 23.4 days in 2020 (PigCHAMP, 2005; PigCHAMP, 2020). Average wean ages from 2005 would be below the range of wean ages in 2020. As wean age in the swine industry is trending to older ages, producers are tasked with determining the best wean age for their system. When considering changing wean age, producers must determine if the costs associated with the change outweigh potential benefits. To better understand the impact of increased wean age, DNA Genetics partnered with Pipestone Applied Research (PAR) to conduct three wean age trials on DNA L600 x DNA L241 animals as well as collected data from Pillen Family Farms production system.

2. Reproduction

As producers consider increasing wean age, there are two options for the sow farm. Producers can add farrowing crates or reduce sow inventory. Depending on building permit availability, costs, and opportunity for market access (California's Proposition 12), both may be viable options. For example, a 2500 head sow farm that would like to increase wean age by 4 days would need to add an additional 60 crates, which would be a $270,000 USD investment (assuming $4,500 USD per crate). Instead of adding additional farrowing crates, the producers could reduce their sow herd inventory. When reducing the sow herd inventory, the producer is also decreasing the total number of pigs weaned while having the same fixed asset costs. To gain an additional 4 days of wean age, the same 2500 head producer would have to decrease the herd by 476 sows or 19% of the herd inventory. This loss of sow inventory is similar to the projected reduction in inventory from a Proposition 12 retrofit (Frobose, 2021). Both options require a significant increase in performance or market access to receive an adequate payback.

Adding an additional 4 days of wean age reduces the number of turns through farrowing crates, thus decreasing the litters per sow per year by 0.065. This decrease in litters per sow per year is offset by an increase in total born (TB) and born alive (BA) in subsequent litters (Figure 1, Figure 2). Total born and BA performance increased by 0.8 when wean age increased by 5 days (Figure 1). This increase in TB is higher than a comparison of DNA L241 farms across production systems, which showed a 0.09 TB increase per day of extra wean age (Figure 2).

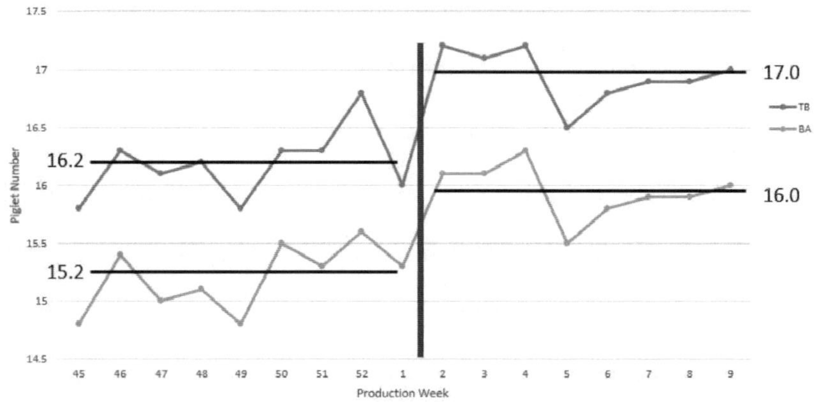

Figure 1. Reproductive performance by production week pre and post wean age change on a 5600 head commercial sow farm, when subsequent litters started farrowing.

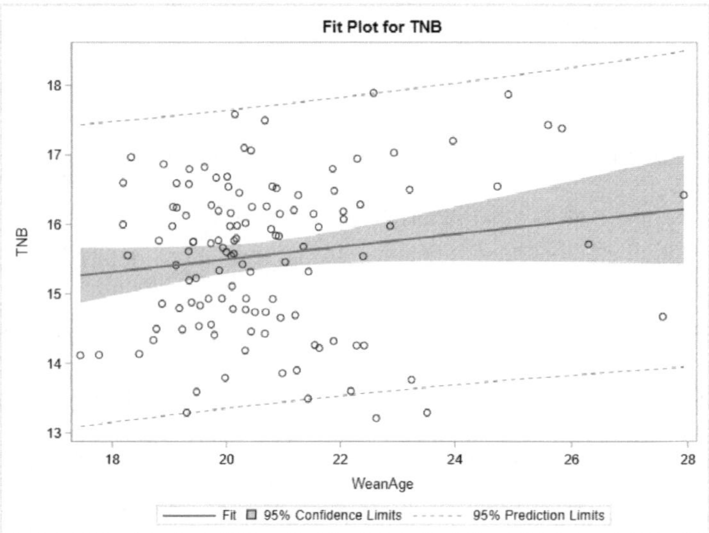

Figure 2. Total number born per average wean age, where each dot represents a unique DNA L241 farm's 52-week summary data. Dataset includes 116 farms representing over 280,000 sows.

Increased TB due to increased wean age could be attributed to sows having a longer lactation length. Towards the end of the lactation period sows are typically maximizing feed intake. Therefore, the extra days during the lactation period provides sows the opportunity to transition from a negative energy balance to a positive energy balance. A 317.5-gram increase in average lactation intake per additional day of wean age was observed from data collected from multiple Pillen sow farms. Increased wean age, not only improved subsequent reproductive performance, but it also improved pig wean weight. In three wean age trials with Pipestone Applied Research (PAR) with approximately 2700 pigs born alive per trial, pigs were 226.8 grams heavier at weaning for every additional day of wean age from 19-26 days of age (PAR 2017; 2019a and 2019b).

3. Grow Finish

Additional wean age has the biggest benefits in grow finish. Therefore, to justify the costs associated with increased wean age, producers need to be a farrow to finish producer or have a large enough weight slide contract to justify the change.

As pigs age during the lactation period, they become more physiologically mature, resulting in an increased ability to eat and digest non-milk/liquid feeds. Due to increased physiological maturity, older wean age pigs tend to grow faster during the nursery period regardless of wean weight. A 4 kg pig at 23 days of age grows the same or slightly better than a 4.5 kg pig at 21 days of age during the nursery phase (Bruns, et al. 2021). Additionally, quality light pigs (<4.5 kg) weighing the same at weaning were 589.7 grams heavier at the end of the nursery phase for each additional day of wean age (PAR, 2019a).

Most of the value to wean age occurs during the nursery phase. The first week of the nursery period is especially critical. Faccine et al. (2020) reported that 35% of pigs weaned at 19-days of age lose weight in the first week post weaning compared with only 12% of the pigs weaned at 25 days of age. In a separate study, pigs that were 26 days of age at weaning were 5.9 kg heavier at day 60 post weaning compared to pigs weaned at 20-days of age (PAR, 2019b). The value of wean age stabilized by day 140 post weaning, where the difference between 20 and 26 day wean ages was 7.23 kg (PAR, 2019b). These results were similar to another trial that compared day 18 and day 25 wean ages. Pigs weaned on day 25 had a 27.2 g ADG wean to finish advantage and a 1.04% increase in full value pigs (PAR, 2017). While there is a growth response to wean age, there is no significant differences in feed conversion based on the wean age of the pigs (PAR, 2017; 2019b).

Taking results to a commercial environment, is the end goal of the research process. Nursery ADG downstream from a 5600 head commercial sow farm increased 9.9 grams per each additional day of wean age (Figure 3). Nursery mortality decreased 0.12% for each day of additional wean age from the same flow (Figure 4). Pigs tend to grow faster in the nursery as wean age increases due to consuming nursery feed quicker post weaning, resulting in an easier pig to start in the nursery.

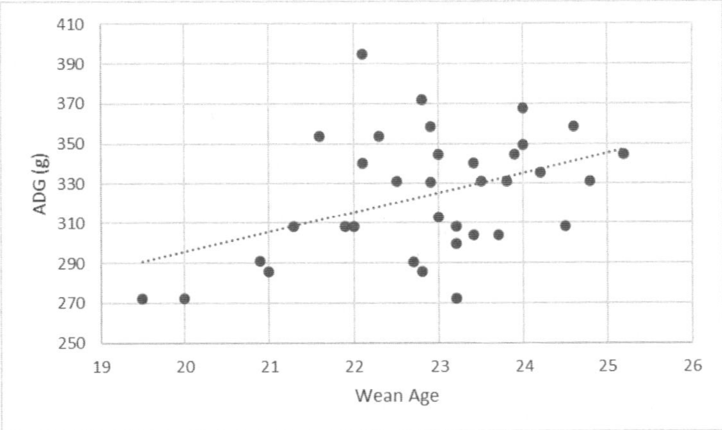

Figure 3. Nursery average daily gain by wean age from a 5600 head commercial sow farm where each dot represents a unique wean group in a 7-week nursery turn.

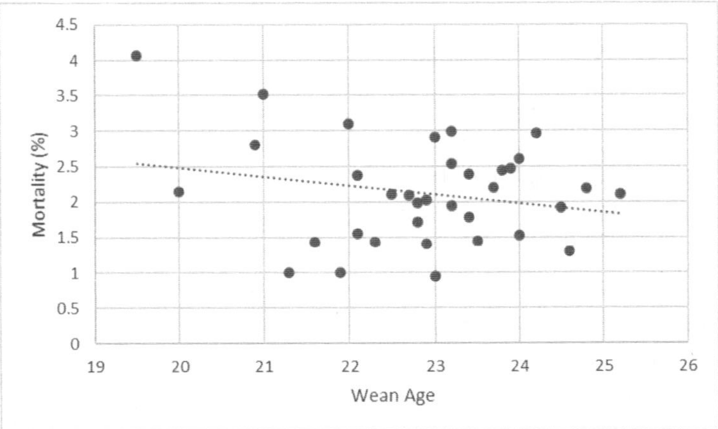

Figure 4. Nursery mortality by wean age from a 5600 head commercial sow farm where each dot represents a unique wean group in a 7-week nursery turn.

Many producers considering converting to Antibiotic Free (ABF) production systems or flows within a production system, investigate adding wean age to reduce the need for intervention strategies. In a PAR research trial with ABF production, the 25-day wean age group had 6.6% more full value (able to receive ABF premium) pigs than pigs weaned on day 18 (Table 1) (PAR, 2017).

Table 1. Antibiotic free (ABF) wean to finish by wean age trial results. Full value pigs were eligible to receive an ABF premium.

ABF Pens	18 Day Wean	25 Day Wean
Pigs	540	540
Received Treatment	19 (3.5%)	23 (4.3%)
Dead or Removed	41 (7.6%)	18 (3.3%)
Light Culls	13 (2.4%)	7 (1.3%)
Full Value	474 (87.8%)	510 (94.4%)

4. Summary

The North American swine industry is moving to an older wean pig due to the advantages seen in the sow farm and grow finish barns. Marketing opportunities (Prop 12, ABF, etc.) are also providing advantages to increased wean age. However, these changes come with an economic and management cost. It is up to each producer and production system to identify the optimum wean age for their respective health status, marketing goal, or management level.

References

Bruns, C., McNeil, B., Sonderman, J. and Rathje, T. (2021). The relationship between early life measures and finishing performance. AASV, Industrial Partners Session 4, 2021.

Faccin, J., Laskoski, F., Hernig, L. F., Kummer, R., Lima, G., Orlando, U., Gonçalves, M., Mellagi, A., Ulguim, R. R., and Bortolozzo, F. P. (2020). Impact of increasing weaning age on pig performance and belly nosing prevalence in a commercial multisite production system. *Journal of animal science*, *98*(4), skaa031.

Frobose, H. (2021). The Gray Areas of California Proposition 12. Gestal Swine Summit.

Pipestone Applied Research (PAR) and DNA Genetics. (2017) Comparison of 18 and 25 day old, weaned pigs growth performance, sows productivity, response to different diet types, and impact of high and low indexed sires. Unpublished Data.

Pipestone Applied Research (PAR) and DNA Genetics. (2019a) Comparison of 20 to 27 day wean age on a simple, standard, and complex diet. Unpublished Data.

Pipestone Applied Research (PAR) and DNA Genetics. (2019b) Wean age regression (20 days to 26 days) x Diet (Simple diet vs. Complex diet). Unpublished Data.

PigCHAMP. (2005). PigCHAMP Benchmarking USA year summary. https://www.pigchamp.com/benchmarking/benchmarking-summaries.

PigCHAMP. (2020). PigCHAMP Benchmarking USA year summary. https://www.pigchamp.com/benchmarking/benchmarking-summaries.

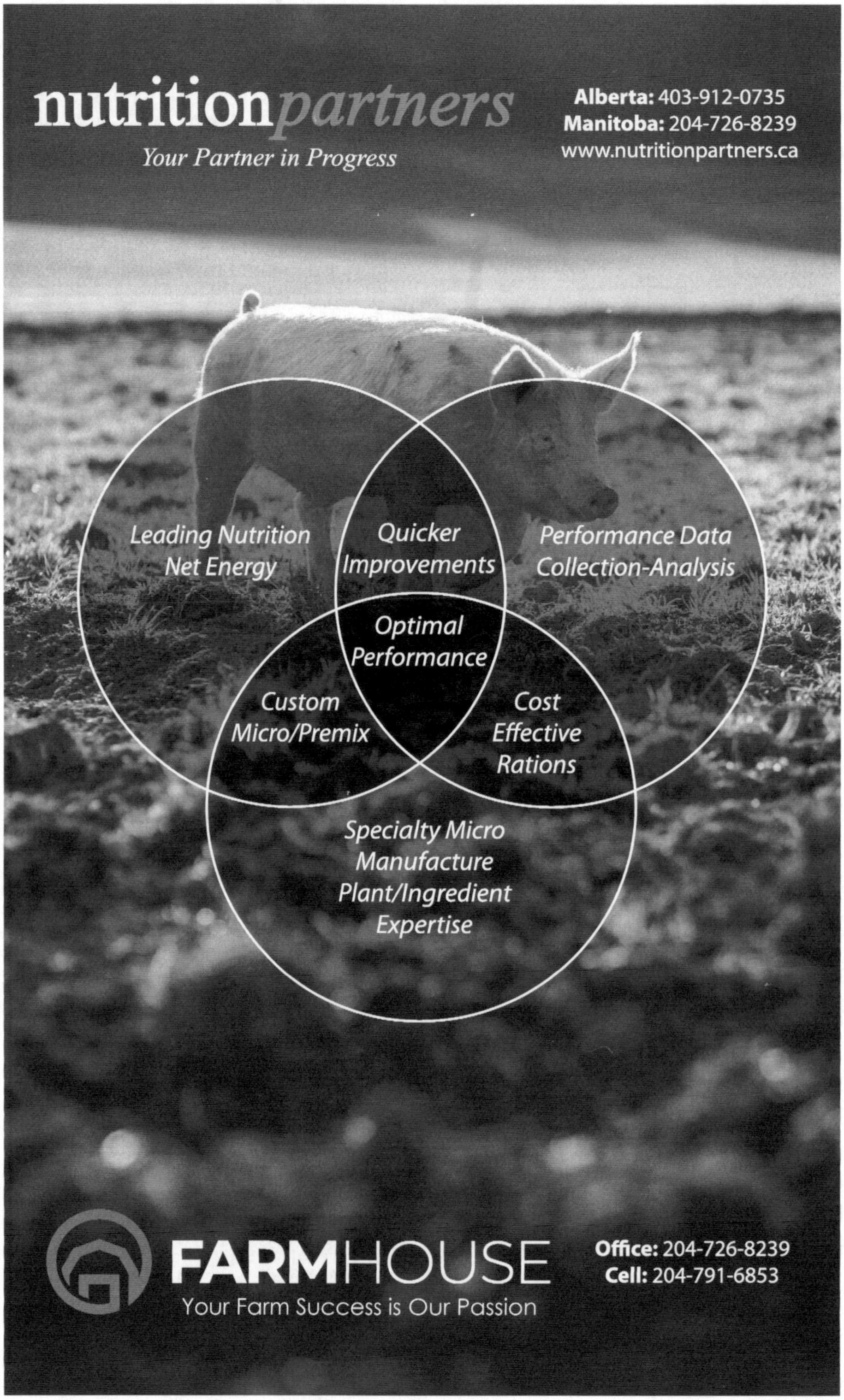
nutritionpartners
Your Partner in Progress
Alberta: 403-912-0735
Manitoba: 204-726-8239
www.nutritionpartners.ca
Leading Nutrition
Net Energy
Quicker
Improvements
Performance Data
Collection-Analysis
Optimal
Performance
Custom
Micro/Premix
Cost
Effective
Rations
Specialty Micro
Manufacture
Plant/Ingredient
Expertise
FARMHOUSE
Your Farm Success is Our Passion
Office: 204-726-8239
Cell: 204-791-6853

Have You Hit a Farrowing Rate of 95% This Year?

Jean-Philippe Martineau & Dan Bussieres

1. Introduction

There is a large variation within producers when it comes to farrowing rates. While some producers maintain a steady rate of 95% throughout their year, others often struggle to surpass even 85%. This article will include studies that have shown increases in litter sizes as farrowing rates and litter sizes are highly correlated (figure 1). It is presumed that the farrowing rates within these studies were also positively affected. In order to optimize sow reproduction, nutrition, sexual hormones, and breeding techniques need to be well managed. The intent of this paper is to share some of the latest in-house peer-reviewed trials that have been found to optimize sow reproduction so that you too can frequently attain a farrowing rate of 95%.

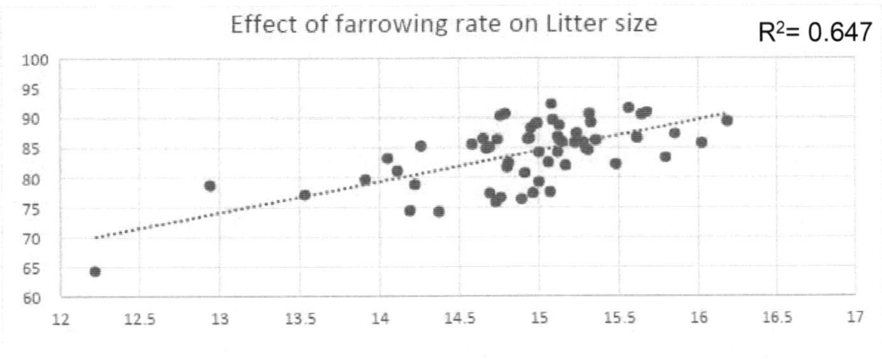

Figure 1: Effect of farrowing rate on litter size (2021 average from 60 farms across Canada)

2. Finding Low Intake Sows During Lactation

The standardized ileal digestibility (SID) lysine intake during lactation is highly correlated to farrowing rates and next litter sizes. The correlation between the average SID lysine intake of sows during a 21-day lactation period and their next litter sizes, based on in-house research, is shown in figure 2. The smallest next litter size was measured from sows that had had an average daily intake of less than 50g of SID lysine/day over a 21-day lactation period. Internal data shows that 10% of sows do not hit that feed intake target when fed a 1% SID lysine ration during lactation.

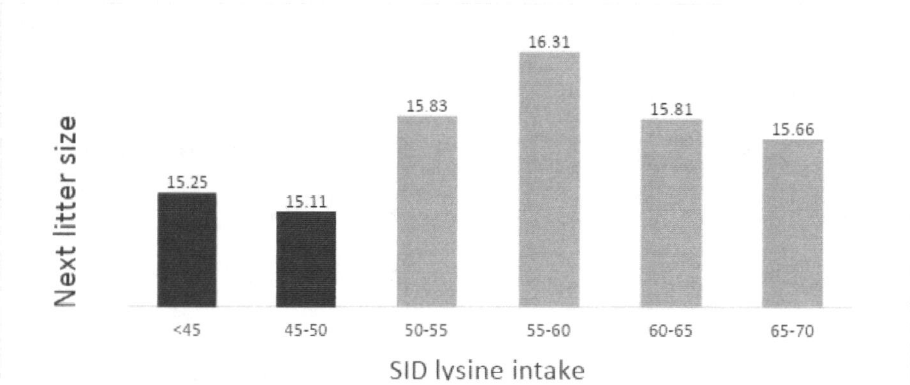

Figure 2: Effect of the average sow SID Lysine intake during a 21 days average lactating period on the next litter size (In-House trial)

In order to help the low feed intake sows, they must first be identified by assessing feed intake during their first 4 days in lactation. Once the 10% of sows with the lowest feed intake are identified, a top dress may be provided. A top dress is a concentrate of energy (fat and oils), protein (soybean meal and synthetic amino-acids) and micronutrients. Providing daily access to a concentrate allows more sows to achieve the SID lysine intake target and in turn improves your herd's average wean-to-estrus interval (WEI), farrowing rate, and litter size.

3. Using the Right Sexual Hormones

Knowing that parity 1(P1) sows will have longer WEIs than parity 2+(P2+) sows due to a SID lysine + energy deficit during lactation periods (figure 3), some producers choose to routinely administer PG600™ (follicle-stimulating hormone [FSH] and luteinizing hormone [LH]) to their P1 sows.

Parity	**WEI (days)**
P1	7,21
P2+	5,72

Figure 3: Effect of parity on Wean to estrus internal (WOI) (in house data 2019)

Based on in-house trials, administering PG600™ to P1 sows is an effective way to get them to show an estrus within 7 days (+5.5%). However, the administration of PG600™ does not improve the overall farrowing rate (-3.71%) and decreases litter size (-0.94 piglets/litter) as shown in figure 4.

Performance parameters	**Control**	**Test**	**P value**
Nb of gilts at first weaning	109	112	
Weaning-Service interval	7.53	6.84	0.4782
Nb sows bred	102	111	
Bred sows (%)	93.58%	99.11%	p<0.05
Nb farrowed	92	96	

Farrowing rate (%)	90.20%	86.49%	NS
Total born/litter	14.72	13.78	0.1048
Bred sows which farrowed (%)	90.20	86.49	NS
Weaned sows which farrowed (%)	84.40	85.71	NS

Figure 4: Effect of PG600™ on farrowing rate and litter size (in house trial 2012)

There is now a new product on the market called MaprelinXP-10™(Perforelin) which appears to increase litter size when administered to P1 sows (figure 5) by targeting the FSH hormone only. By doing so, it increases the number of follicles that achieve the size needed to be fertile. Based on our research, MaprelinXP-10™ is superior to PG600™ for boosting sow performance through the use of a sexual hormone.

	Control	**Maprelin™ (2cc)**
Total Born	14,86	15,31

Figure 5: Effect of Maprelin™ on litter size in P1 sows

Another sexual hormone that can be administered is altrenogest, commonly known as Regu-MATE™. In a recently completed trial, performance between naturally cycling gilts and gilts cycled using altrenogest were compared. Ensuring that first litter performance is not compromised by altrenogest was important as several 4-week batch farrowing barns cycle their gilts using this hormone. This trial validated that the use of altrenogest poses no significant effect on litter size (figure 6).

	Control	Altrenogest	SEM	Pvalue
# sows	95	110		
TB	14,04	13,87	0,3558	0,7447
BA	13,47	13,14	0,3498	0,5208
Average birth weight	1,44	1,47	0,022	0,4174

Figure 6: Effect of cycling gilts with an altrenogest on litter size

4. Breeding Techniques and Strategies

Good breeding techniques start with proper management of gilts by ensuring they are fully fed until breeding and bred at both the target age (200 days) and the target weight (145kg-can vary depending on genetic line). Starting boar exposure at 150 days will allow for more gilts to reach puberty by the age of 180 days (Roongsitthichai et al. 2014).

Gilts should be bred using the regular artificial insemination (AI) technique while P1+ sows should be bred using the post-cervical artificial insemination (PCAI) technique in order to optimize litter size as shown in figure 7.

Parameter	**Standard AI**	**PCAI**	**P-Value**
# of sows	468	444	
Farrowing rate, %	89.93	89.25	>0.10
Total born, pigs/litter	14.65	15.03	0.0776

Figure 7: Effect of breeding with PCAI on sow performance (in house trial 2014)

When using the AI technique, 75ml semen bottles containing 2 billion sperms are used while the PCAI technique can be accomplished with 56ml semen bottles containing only 1.5 billion sperms. Despite the PCAI technique using 25% less sperm cells per semen bottle, a recent in-house trial showed that performance was not compromised (figure 8). As a result, less boars are needed for the same quantity of bred sows. This increases the economic genetic potential of the pool of boars needed.

Parameter	2.00 Billion	1.50 Billion	P-value
# of sows	728	731	
Farrowing rate, %	92.04	90.36	>0.10
Total born, pigs/litter	15.38	15.46	0.6423

Figure 8: Effect of semen concentration on sow performance using the PCAI breeding technique

Another factor considered to optimize performance is the use of the right semen extender. Not all semen extenders are created equally. The success of semen extenders can vary depending on the sperm concentration used and the number of days before use. Therefore, some extenders might better fit your needs than others. Figure 9 shows results from a semen extender trial where different extenders showed similar farrowing rates within treatments but significantly different litter sizes (-0.57 piglet/litter).

Parameter	Extender A	Extender B	P-value
# of sows	624	600	
Farrowing rate, %	88.80	89.80	0.5559
Total born, pigs/litter	**15.30**	**14.73**	**0.0043**

Figure 9: Effect of semen extender on performance (in-house trial 2018)

5. Summary

Different strategies aimed at improving farrowing rates and litter sizes were presented in this paper. It is important to validate these strategies with your veterinarians as some of the products studied require prescriptions. If you have not been hitting a farrowing rate of 95% and your litter size is not where you would like it to be, do not hesitate to speak to your consulting team because several solutions exist to help you reach those targets.

References

Roongsitthichai A., Olanratmanee E, Tummaruk P. 2014. Influence of growth rate and onset of boar contact on puberty attainment of replacement gilts raised in Thailand.Tropical Animal Health and Production v46 n7 (201410): 1243-1248

Nutrition as it Relates to Increased Sow Productivity: The value of fat in lactation diets

E. K. Holmes, Ph.D.

Cargill Animal Nutrition, Lewisburg, Ohio

1. Introduction

The lactation period is an energetically demanding phase of the sow's life and energy is the largest nutritional cost of a diet. Even so, sows continue to produce more and more pigs as genetics improve, placing a greater energetic demand on the sow during lactation. Energy can come from fat, carbohydrates, and protein, but it has long been thought that a certain portion or a minimum amount of the energy in a lactation diet must come from fat. Since the beginning of 2020, prices of fats and oils have risen dramatically, in some instances, at least double the typical price (The Jacobsen, 2021). This makes any amount of added fat in the diet a more expensive proposition. This has led producers to ask whether a lactation diet can be formulated without fat. The objectives of this paper are to discuss: 1) the causes for the dramatic price increase of fats and oils; 2) potential outlook for the fat market; 3) nutritional attributes of fat; and 4) potential approaches to meet the energy requirement of the lactating sow with less added fat.

2. Fat Market

The price of fat has increased exponentially since early 2020 (The Jacobsen, 2021). This can be attributed to technology advancements of renewable diesel and associated supply and demand. Renewable diesel is produced from fats, oils, and grease feedstocks. Because of the chemistry of renewable diesel, it can be used at 100% replacement to traditional diesel refined from crude oil, compared with biodiesel which can only be blended up to 20% with traditional diesel. In the U.S., if all intended renewable diesel expansion projects come to pass, production capacity could reach five billion gallons per year by 2024 (EIA, 2021 and JPM, 2021), up from 600 million gallons per year at the end of 2020 (EIA, 2021). As renewable diesel production has increased, so has the demand and price for these feedstocks, and thus, the livestock feed industry is feeling the competitive pressures. Some forecasts estimate that there may not be enough fats and oils to keep up with the demand for renewable diesel, let alone enough fats and oils to use in livestock feeds. Importing fats and oils for either renewable diesel production or livestock diets is an option, albeit at a cost. Therefore, sourcing fat may become challenging, or at best it will continue to be an expensive feed ingredient to use in swine diets.

3. Fat as an Ingredient

Fat is found in all major feedstuffs at various concentrations. Primary examples of fat and oil sources used in swine diets include corn oil, choice white grease, and animal-vegetable blends. Each of these sources has different physical and molecular properties that offer advantages and disadvantages to their use. These fats and oils are a very concentrated source of energy and have a lower heat increment making the pig's metabolism of fats and oils more efficient than carbohydrates or proteins. Additionally, fats and oils provide essential fatty acids to the diet, facilitate absorption of fat-soluble vitamins, and at low inclusion rates have traditionally been used for dust control (NRC, 2012). Fats and oils also can affect feed flowability; too much fat can cause flowability issues of meal feed.

As research has progressed, nutritionist have learned more about the functional properties of fats and oils and how they can affect animal performance. Fats and oils are comprised of different subunits called fatty acids. These fatty acids contribute to the quality of the fat and determine how sensitive the fat source will be to peroxidation and rancidity. Feeding peroxidized oils can lead to cell damage known as oxidative stress. Rancidity produces off flavors that could reduce palatability of feed and suppress feed intake (Wealleans et al. 2021). In the lactating sow this would be undesirable as the goal of feeding the lactating sow is for milk production and litter growth while minimizing body condition loss. This is achieved with high feed intake by the sow during lactation. Su and others (2017) reported reduced feed intake in lactating sows when sows were fed peroxidized oil, but did not observe a difference in performance. In a meta-analysis, Hung et al. (2017) reported that feeding peroxidized oils to swine and poultry reduced growth performance.

Another important property of fats and oils is fatty acid contribution and intake. Linoleic and linolenic acids are essential fatty acids that cannot be made by the body and must be provided through the diet. Linoleic and linolenic acids are components of cell membranes, cell signaling, and are the precursors to making other longer fatty acids that are important for organ development and a host of biological activities (Lin et al. 2013). The fatty acid profile in tissues of the suckling piglet will reflect that of the diet the sow is consuming (Lauridsen and Jensen, 2007). Linoleic acid is one of the primary fatty acids secreted through the milk (Lauridsen and Jensen 2007) and has been shown to improve reproductive efficiency and sow retention (Rosero et al., 2016). In a review of 12 studies, Rosero and co-workers (2016) postulated that the greatest benefit of supplementing fat to lactating sows are the essential fatty acids. These authors proposed feeding a minimum of 100 g/d of linoleic acid to prevent negative balance or 120 g/d to maximize sow reproductive efficiency (farrowing rates and subsequent litter size) (Rosero, 2014).

4. Methods to meet energy and essential fatty acid requirement

4.1 Feed Intake

Nutrient intake during lactation is a key component to allow the sow to achieve her genetic potential for maximizing milk production while maintaining body condition. Daily energy intake is a function of both feed intake and diet concentration. Diets are formulated to meet daily nutrient intake, so the nutrient density or concentration will reflect the sows expected feed intake; a lower feed intake will require a more nutrient dense diet. The modern sow is leaner and has a lower appetite. Feed intake is commonly compromised during periods of heat stress, disease challenge, and younger parity sows. Encouraging higher feed intake is one way to lower the nutrient or energy density of the diet, which will lower the need for more concentrated sources of energy (e.g. fat).

Several variables can influence voluntary feed intake, and some can be managed at the barn level and some through diet formulation to encourage high voluntary feed intake in lactating sows. Feeding higher fiber diets in gestation helps sows transition better from a gestation diet to a lactation diet by increasing the capacity of the gastrointestinal tract before lactation when the sow will be consuming more feed. Constipation can occur due to the limited feed intake during gestation or simply changing between gestation and lactation diets (Peltoniemi and Oliviero, 2015). This can be alleviated through ingredient management. Particularly during warm summer months, feeding times may need to be adjusted to accommodate times when sows are more likely to eat, either in the early or later parts of the day. Keeping sows cool during summer heat is especially imperative, so appropriate ventilation and adequate water intake are critical. Having sows in proper body condition coming into farrowing is also important; heavy conditioned sows have poorer feed intake during lactation. This list is not fully exhaustive but highlights key factors to keep in mind when striving for more lactation feed intake and lower nutrient dense diets that use less added fat.

4.2 Nutrient utilization

The sow has a larger degree of fermentation capacity compared to nursery or grow-finish pigs (Noblet and Le Goff, 2001). Non-starch polysaccharides are more difficult for the sow to utilize relative to other carbohydrates. Utilization of these nutrients would facilitate capturing energy that would otherwise be wasted or excreted. Exogenous enzymes give pork producers a means of being able to achieve enhanced nutrient digestion. Xylanase is a carbohydrase enzyme that breaks down arabinoxylans, found in plant cell walls, to nutrients that are readily metabolized or fermented to yield energy. Xylanases may have other important metabolic functions such as a prebiotic or stimbiotic effect while supporting gut barrier function. Cereal grains (e.g., corn and wheat) and corn-based coproducts such as dried distillers grains with solubles (DDGS) have an abundance of arabinoxylans. The effect of xylanase in wheat-based swine diets has been moderately consistent and in corn-based diets more variable (Petry and Patience, 2020). A review of eight studies where xylanase was supplemented to sows reported reduced body weight loss during lactation and heavier litter weight gain as a result of higher energy digestibility (Cozannet et al., 2018). Research by Petry and Patience (2020) has advanced the swine industry's understanding of the mode of action of xylanase in diets containing fiber from corn and corn coproducts and has helped to pave a path for implementation of xylanase as a tool to capture more energy from the diet.

Many lactation diets contain either corn or wheat, while fewer contain the corn coproduct DDGS. Corn DDGS is a suitable energy source, but the energy will come more from fat, crude protein, and fiber, and less from simple carbohydrates

(Petry and Patience, 2020). Research by Greiner and co-workers (2015) demonstrated that DDGS can be included in lactation diets up to 30% without adversely affecting feed intake or sow and litter performance. However, feeding diets containing 40% or 50% DDGS to lactating sows may reduce feed intake and litter performance (Greiner et al., 2015). Along with providing energy, the corn oil in corn-DDGS would contribute a favorable level of linoleic acid to lactation diets (NRC, 2012). Moreover, corn DDGS have a relatively high concentration of arabinoxylans, the main substrate of focus for the use of xylanase (Petry and Patience, 2020). Other high energy ingredient sources such bakery byproduct could be considered to provide lactating sows energy while limiting the amount of fat needed in the diet.

The objective discussed herein was to consider how lactation diets could be approached with limited added fat. Fat can be added to diets and provides a dense energy source. However, current and future prices of fats and oils mean added fat adds considerable cost to the diet. Yet, the sow will only have an increasing need for energy as sows produce more pigs. Focusing on sow lactation feed intake, ingredient selection, and improving nutrient utilization will facilitate curbing the cost of production while meeting the nutritional needs of the sow and her litter.

References

Cozannet, P., P.G. Lawlor, P. Leterme, E. Devillard, P. Geraert, F. Rouffineau, and A. Preynat. 2018. J. Anim. Sci. 96: 2777-2788.

Greiner, L., C. Neill, G. L. Allee, X. Wang, J. Connor, K. Touchette, and J. L. Usry. 2015. The feeding of dried distillers' grains with solubles to lactating sows. J. Anim. Sci. 93:5718-5724.

Gresh. P, J. M. Royall, N. J. Lampman, T. Palmer. 2021. Renewable Diesel: Industry Primer and Multi-Year Outlook. J.P. Morgan North America Equity Research.

Lauridsen, C., and S. K. Jensen. 2007. Lipid composition of lactational diets influences the fatty acid profile of the progeny before and after suckling. Animal. 1:952-962.

Lin, X., M. Azain, J. Odle. 2013. Lipids and lipid utilization in swine. In: Sustainable swine nutrition. Blackwell Publishing Ltd., Oxford, UK. p.59-79.

Noblet, J., and G. Le Goff. 2001. Effect of dietary fibre on the energy value of feeds for pigs. Anim. Feed. Sci. and Tech. 90: 35-52.

NRC. 2012. Nutrient Requirements of Swine. 11th ed. NRC Press.

Petry, A., and J. F. Patience. 2020. Xylanase supplementation in corn-based swine diets: a review with emphasis on potential mechanisms of action. J. Anim. Sci. 98:1-12.

Peltoniemi, O.A.T., and C. Oliviero. 2015. Housing, management and environment during farrowing and early lactation. In: C. Farmer (ed.) The Gestating and Lactating Sow. p.231-252. Wageningen Academic Publishers.

Rosero, D. S., R. Dean Boyd, J. Odle, E. van Heugten. 2016. Optimizing dietary lipid use to improve essential fatty acid status and reproductive performance of the modern lactating sow: a review. J. Anim. Sci. Bio. 7:34.

Rosero, D. S. 2014. Nutritional Value of Dietary Lipids in Lactating Sow Diets. [Doctoral dissertation, North Carolina State University].

The Jacobsen. Animal Fats and Oils. https://thejacobsen.com/price-reporting/animal-fats-oils/ Accessed September 30, 2021.

U.S. Energy Information Administration. 29 Jul, 2021. U.S. renewable diesel capacity could increase due to announced and developing projects. https://www.eia.gov/todayinenergy/detail.php?id=48916

Su, G., J. Zhao, G. Luo, Y. Xuan, Z. Fang, Y. Lin, S. Xu, D. Wu, J. He, and L. Che. 2017. Effects of oil quality and antioxidant supplementation on sow performance, milk composition and oxidative status in serum and placenta. Lipids in Health and Disease. 16:107.

Wealleans, A. L., K. Bierinckx, M. di Benedetto. 2021. Fats and oils in pig nutrition: Factors affecting digestion and utilization. Ani. Feed. Sci. Tech. 277.

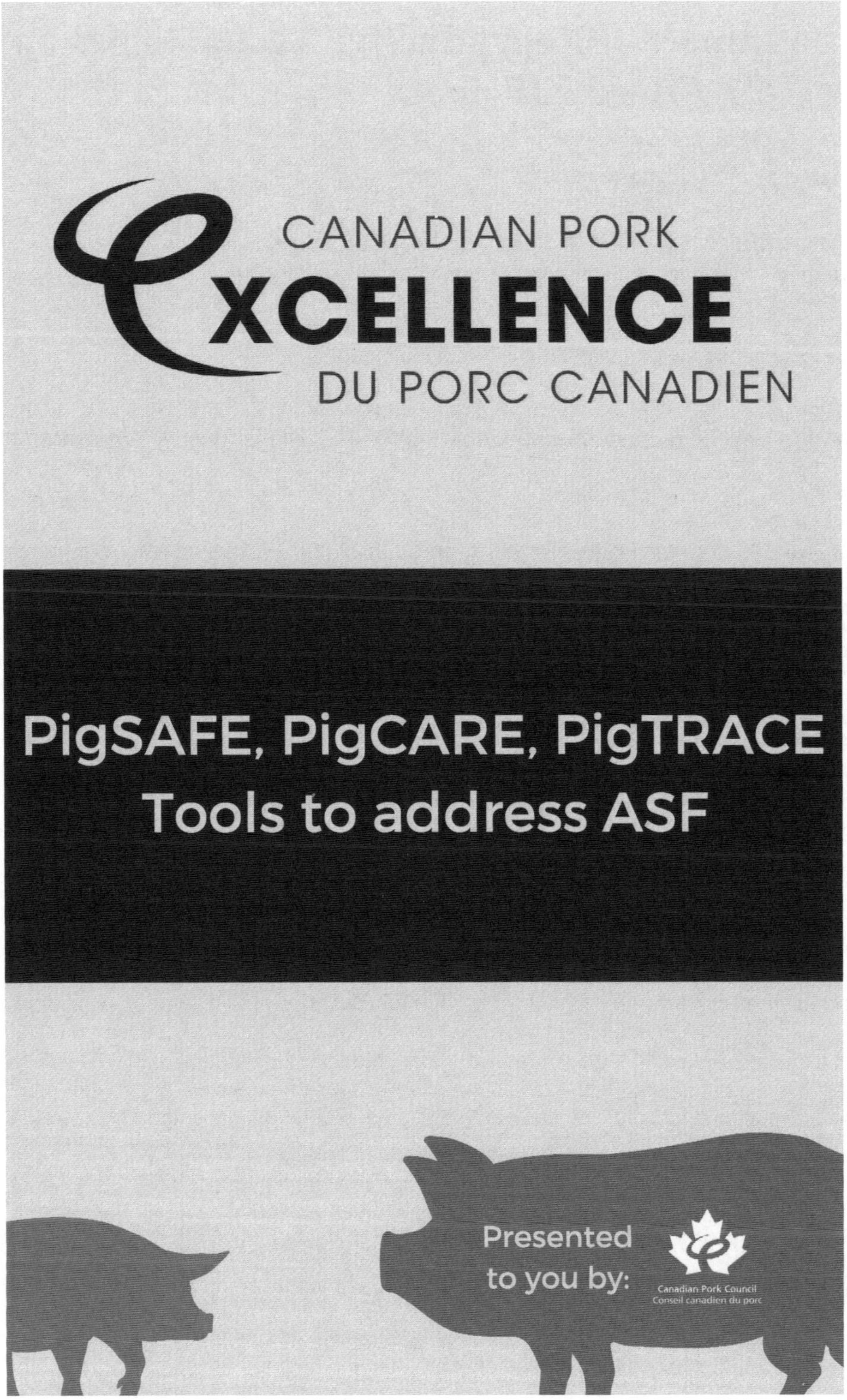

CANADIAN PORK
EXCELLENCE
DU PORC CANADIEN
PigSAFE, PigCARE, PigTRACE
Tools to address ASF
Presented
to you by:
Canadian Pork Council
Conseil canadien du porc

The Impact of Reproductive Technologies on Breeding Herd Efficiency

Kilby L. Willenburg

Fast Genetics, 8137 Forsythia St. Middleton, WI 53562, US
Email: kwillenburg@fastgenetics.com

Introduction

The last few years have been a whirlwind for the swine industry. What seems to be daily market fluctuations in 2021 with slaughter prices reaching levels comparable to the PED inflicted market of 2014. Prices at the counter are squeezing producers and consumers alike, which is perplexing given the record number of pigs slaughtered particularly in the first quarter of 2021. Producers are expected to be profitable in 2022 despite an increase in production costs (where the price of corn has more than doubled from a year ago), threat of inflation and looming presence from a likely increase in cost of living. The pandemic has undoubtedly played a role in animal agriculture with labor shortages being partially to blame for the increase in consumer costs. Fear of COVID, unemployment benefits and government assistance have all contributed to the labor crisis forcing farms to compete with local restaurants and retailers. Production systems will have to be creative and embrace new technologies and leverage employees to fill this void to maintain current production levels. In addition, there is a need for significant increases in production across all agriculture sectors as the world population is expected to rise to 10 billion by 2050. The world will require a lot more food and farmers will face pressure to keep up with demand with minimal increases to existing infrastructure, which will have to be addressed at all levels of production with technology. This paper will focus on improving breeding herd efficiencies with reproductive and non-reproductive technologies with emphasis on the need for precision agricultural strategies.

Artificial Insemination

The swine industry has been driven to improve reproductive efficiency with the caveat of not hindering fertility. Artificial insemination (AI) has been instrumental in improving reproductive efficiency by reducing labor and time required for heat detection and insemination (22 min vs 6 min) compared to natural mating (Flowers and Alhusen, 1992, Knox et al. 2013), genetic improvement, and enhanced biosecurity and disease control (review by Knox, 2016). Artificial insemination is used in more than 90% of farms around the world (Riesenbeck, 2011) and is arguably one of the greatest successes in reproductive technology adoption. Cervical artificial insemination (CAI) has been the most predominant breeding method in the swine industry for several decades (Knox et al. 2013) due to the inexpensive cost of the catheter and relative ease of training and use with doses containing 2.5-3 billion cells in 75 to 100 mL (Knox, 2016). However, lowering the sperm cell number per dose and overcoming technician variation and technician fatigue has supported the higher cell number per dose. Post cervical artificial insemination (PCAI) or intrauterine insemination (IUI) has garnered significant attention because it requires less time to perform compared to CAI (7-10 min vs 1-2 min) (Review by Kraeling and Webel, 2015), uses fewer cells (1-2 billion cells), and smaller volumes (e.g. 45 mL) with no reduction in fertility (Sbardella et al. 2014). The downside to PCAI has been in broad application use in gilts and primiparous sows due to size of the opening of the cranial end of the cervix. Deep uterine insemination (DUI) is another option, albeit limited in use, for inseminating reduced numbers of sperm compared to CAI and PCAI. In hormonally synchronized sows Vazquez et al., (2005) inseminated $50x10^6$, $200x10^6$, $1x10^9$ cells per dose and reported farrowing rates of 92.3, 88.9, 88.6%, respectively compared to an 87.5% farrowing rate for a 3 billion cell, non-synchronized control. The litters sizes for each treatment were 9.41, 9.75, 9.61 and 10.02, respectively. Furthermore, in a field trial (Belstra et al., 2020) with hormonally synchronized sows (Table 1), there were no differences ($P > 0.5$) in pregnancy rates from a semen titration study using a single fixed timed AI with either DUI or PCAI with sperm levels ranging from 1.2 billion to 75 million. However, this trial did show a gradual decline in total number of embryos collected as the sperm dose concentration decreased (16.1 ± 0.7 to 11.5 ± 0.7, $P < 0.05$). A DUI catheter, which is longer than 1.5 m bypasses the cervical folds and manipulates the length and coiled nature of the uterine horn prior to depositing cells in one of the uterine horns (Vazquez et al., 2005). Despite sufficient published data, appeal has been unpopular as this catheter is not cost effective and more suited for technologies like frozen-thawed or sexed sorted semen. Finally, as a bridge between CAI and PCAI in gilts and lower parity animals, a device has been developed for deep cervical semen deposition (Dp CAI). There were no differences ($P > 0.05$) in total born (13.7 ± 2.6 vs 13.1 ± 3.5) and farrowing rate (83.6 vs 87.5) for gilts bred with CAI (n = 136, 2.5×10^9 sperm cells) or with Dp-CAI (n = 1166, 1.5×10^9 sperm cells), respectively (Llamas-López et al., 2019). However, additional studies are warranted comparing balanced sperm numbers across insemination type to fully endorse this AI method.

Table 1. Reproductive performance at 27 d of gestation of sows administered OvuGel[1] at 80 h postweaning and receiving a single fixed-time artificial insemination (AI) at 106 h postweaning containing different numbers of sperm per dose and insemination into the uterine body (IUI) or a deep uterine insemination into a single uterine horn (DUI).

	Insemination Treatment							
Sperm/dose, $\times 10^6$	1,200	600	600	300	300	150	150	75
Semen insemination site	IUI	IUI	DUI	IUI	DUI	IUI	DUI	DUI
Sows, n	67	72	65	65	71	67	65	62
Pregnancy rate, %	80.3 ± 4.9	80.9 ± 4.8	76.2 ± 5.4	77.0 ± 5.4	84.3 ± 4.4	72.3 ± 5.6	84.6 ± 4.5	70.5 ± 5.9
Total number of embryos	16.1 ± 0.7^{a}	15.5 ± 0.6^{ac}	14.8 ± 0.7^{ad}	12.8 ± 0.7^{bcd}	13.5 ± 0.7^{bcd}	12.2 ± 0.7^{b}	12.7 ± 0.7^{bd}	11.5 ± 0.7^{b}
Viable embryos	14.6 ± 0.7^{a}	13.7 ± 0.5^{ac}	13.1 ± 0.6^{ad}	11.5 ± 0.7^{bcd}	11.9 ± 0.7^{bcd}	10.7 ± 0.6^{bd}	11.1 ± 0.6^{bd}	10.3 ± 0.7^{b}

[a-e]Means with different superscripts within a row differ ($P \leq .05$) when all 8 treatment combinations were analyzed.

Hormonal Control Of Estrus And Ovulation

The drive to further improve efficiencies in artificial insemination is evident but gains have been minimal as producers still utilize multiple inseminations per estrus with doses containing several billion cells with a cervical or intrauterine catheter (Knox, 2016). A reduction in the number of inseminations per estrus female and lowering the number of sperm cells per insemination is of great importance to the swine industry and should be practical to implement with current methodologies for ovulation induction in gilts and weaned sows. Ovulation induction has been achieved with hCG (Dziuk and Gehlbach, 1966), pLH (Cassar et al.,2005, Bennett-Steward et al., 2007,) and GnRH agonists (Knox et al., 2017AB, Ulgium et al., 2018) with ovulation occurring between 43-53 h after hCG (Bergfeld et al., 1976), 34-42 h post pLH (Cassar et al.,2005) and 31-39 h for synthetic GnRH (Brüssow, et al. 1990). Decreasing variation in the timing of ovulation with induction methods enables producers to maintain current fecundity levels and creates a path for a single fixed-time AI, which allows for genetic leverage, or utilization of technologies such as sexed sorted or frozen thawed sperm.

Altrenogest (i.e. MATRIX®), an orally active synthetic progestin is commonly used for synchronization of estrus in mature gilts by suppressing pituitary gonadotropins for 14-20 days allowing the corpus luteum (CL) to regress and prohibit growth of new follicles (Review by Kraeling and Webel, 2015). Albeit off-label, several studies have reported an improvement in reproductive performance after altrenogest was given post-weaning (van Leeuwen et al., 2011, Fernandez et al., 2005) particularly in primiparous sows where the delayed returned to estrus is beneficial for recovering lost body condition from lactation (Review by Kraeling and Webel, 2015). Altrenogest administration post-weaning has also been used for batch farrowing production systems to synchronize breeding groups and to reestablish breeding groups following early piglet death from diseases such as PRRS (Review by Kraeling and Webel, 2015). New applications for this old hormone have given producers additional tools to account for inherent variation in the sow herd as well as unsuspecting disease outbreaks.

Artificial Intelligence in Artificial Insemination

A successful ovulation induction program poses challenges to farms due to normal variation in lactation length, parity structure, season and management practices, which can influence post-weaning follicle growth and maturity and overall efficacy of hormonal treatment (Knox and Rodriquez, 2001, Lucy et al., 2001). An alternative method to controlling time of ovulation is a camera-based technology called PigWatch® that uses artificial intelligence (AI) to monitor sow activity post-weaning and notifies the farm at the optimal time for insemination. Data has shown that when used in conjunction with heat detection, PigWatch can identify estrus in 99.78% of sows, but is significantly less (79.46%) without estrus detection (Labrecque and Rivest, 2018). The authors also reported an average of 1.32 inseminations per sow with an 89.52% conception rate (Labrecque and Rivest, 2018). In a related project comparing data from Pigwatch and time of ovulation based on hormonal profiling during estrus, 95% of the inseminations were in the range of 32 hours before ovulation to 8 hours after ovulation with an average fertility rate of 95.1% and 15.26 total born (Klopfenstein et al., 2016). A similar behavior monitoring technology for sows in group housing was designed by Nedap Livestock Management, which is an electronic estrus detection system that records the frequency of female visits to the boar area and notifies the farm when the threshold for estrus has been reached. These smart devices could be an attractive option to overcome labor shortage or reduce the number of doses inseminated per sow but are not likely to predict the timing of ovulation with enough accuracy to facilitate lowering the sperm numbers in inseminations or with frozen and sexed sorted sperm. Another sensor-based technology called the E Doctor and loosely named "Fitbit for Pigs" is being developed by SmartAHC from Singapore but appears to be in the proof-of-concept stage of testing. This wireless ear tag collects real-time data on body temp, physical activities and estrus. Although in its infancy, wearable smart devices have the potential to revolutionize the industry, reduce human error and improve farm efficiencies. This is a relatively new market but according to IDTechEX, a market research firm, the global market for wearable technologies for animals, livestock included, is estimated at $2.6 billion by 2025.

Robotics

Robotics may not directly impact breeding efficiencies, but they do have the ability to automate the slow, monotonous tasks for farmers allowing them to focus on essential, time-sensitive farm responsibilities. The BoarBot, a self-propelled, steerable machine that tethers a boar for performing estrus detection has improved breeding efficiency and has become a cornerstone for estrus detection on many levels as detecting heat prior to this technology required multiple people for simultaneous movement of the boar and technician-induced back pressure. The previous, non-mechanized system was slow, dangerous for the employee that handled the boar and ensuring optimal boar contact was never guaranteed. The BoarBot has reduced the amount of employee injuries and number of people required for heat detection all while maximizing boar contact at the discretion of the operator as well as increase farrowing rates by 2-10% according to the company website. Another product offered by Swine Robotics, Inc is a robotic power washer, which according to the company reduces wash time by 85-90% and worker exposure to disinfectant chemicals. These types of jobs that require minimal training are better served with robotics allowing farm personnel to concentrate on different tasks. Lastly, a feed dispensing, classical music playing robot has become quite popular in Brazil where they are also experiencing a shortage of labor. RoboAgro, the manufacturer, has stated that the robot can save an estimated $8,000 per year for every 1,000 animals by dispensing the exact amount of feed required per animal while reducing employee labor. Collectively, these products give swine farms the tools to mitigate fluctuations in daily routines from employee turnover or health so focus can be directed towards pertinent farm responsibilities such as mating, gilt development and postnatal care in farrowing.

Sexed Sorted Sperm

Flow cytometric sorting is well established in other livestock industries, particularly dairy when in 2016, approximately 5 million straws of sexed semen were processed in the US (Moore and Hasler, 2017). Inguran Sexing Technologies, Inc has been the driver in SSS and since its inception has produced over 110 million straws of sexed sorted sperm via its 26 sorting labs worldwide and is responsible for more than 50 million replacement heifers. In the swine industry, sexed sorted semen is not available for commercial use, yet. However, within Fast Genetics, the application of sexed sorted sperm has created economic value at the genetic nucleus and multiplier farms. At the genetic nucleus level, a faster rate of genetic improvement can be made within specific lines by increasing the selection intensity, whereas at the multiplier level, doubling the output of the preferred gender without impacting infrastructure cost is highly attractive, with potential for improvement in other efficiencies. Commercially, the value of sexed sorted sperm has been estimated at $5-7 per market hog but will vary depending on the production system and desires of the consumer.

Conclusions

As an industry, we have to improve efficiencies at all levels of production and embrace technologies that can replace the mundane, monotonous tasks to focus our attention on breeding, gilt development, and early post-natal care, to name a few. A 5% drop in preweaning mortality in the U.S. would result in over 2 more PSY and enable the swine industry to drop sow inventory by almost 7% to maintain the current level of production. Furthermore, lowering the number of sperm per insemination across the industry to 1 billion and 500 million cells from the standard 3 billion cells could trim boar inventory by 67 and 83%, respectively, assuming 2.2 breeding doses per sow per litter from an average ejaculate with 60 billion cells. Even at 3 billion cells per dose with a single fixed-time insemination there would be 50% fewer boars required to support the U.S. industry. However, the true value for lowering the cell number with fixed timed AI comes with utilizing superior males as shown in Table 2 that compares value added from different boar indexes (125, 150 and 175) leveraged over multiple seminal doses (3, 1 and 0.5×10^9 cells/dose) from double and fixed timed inseminations. Historically, the swine industry has been slow to adapt technology partly because there has not been a driver that offsets the potential loss in fertility, but as labor shortages remain imminent and world population expands companies will have to integrate technologies to improve efficiency and remain competitive during times with high input cost and irregular markets.

Table 2. Proposed model for leveraging index value across multiple semen dose parameters with double and single fixed timed inseminations.

	Boar Leverage ($60x10^9$ Cells per collection)					
Cells/Dose	**$3x10^9$**	**$3x10^9$**	**$1x10^9$**	**$1x10^9$**	**$500x10^6$**	**$500x10^6$**
Collections/Yr	75	75	75	75	75	75
Doses/Collection	20	20	60	60	120	120
Doses/Yr	1,500	1,500	4,500	4,500	9,000	9,000
AI/Estrus	2	1	2	1	2	1
Farrowing Rate (%)	90	88	90	85	87	83
Sows/Yr	750	1,500	2,250	4,500	4,500	9,000
Weaned Pigs/Yr	7,945	15,536	23,834	45,020	42,165	80,452
Slaughter Pigs/Yr	7,468	14,604	22,404	42,319	39,635	75,625

	Index ($0.05/Index point)		
Slaughter pigs/yr	**125**	**150**	**175**
7,468	$0	$9,335	$18,670
14,604	$0	$9,255	$27,510
22,404	$0	$28,005	$56,010
42,319	$0	$25,899	$78,798
39,635	$0	$49,544	$99,088
75,625	$0	$40,531	$135,062

References

Flowers, W.L. and Alhusen, H.D. (1992) Reproductive performance and estimates of labor requirements associated with combinations of artificial insemination and natural service in swine. J. Anim. Sci.70:615–21.

Knox, R.V. et al. (2013) An analysis of survey data by size of the breeding herd for the reproductive management practices of North American sow farms. J. Anim Sci. 91:433–45.

Knox, R.V. (2016) Artificial insemination in pigs today. Theriogenology 85 (1):83-93.

Riesenbeck, A. (2011) Review on international trade with boar semen. Reprod. Domest. Anim. 46 (Suppl. 2):1e3.

Kraeling, R.R. and Webel, S.K. (2015) Current strategies for reproductive management of gilts and sows in North America. J. Anim. Sci. Bio. 6(3):1-14.

Sbardella, P.E. et al. (2014) The post-cervical insemination does not impair the reproductive performance of primiparous sows. Reprod. Domest. Anim. 49:59–64.

Vazquez, J.M. et al. (2005) Inproving the efficiency of sperm technologies in pigs: the value of deep uterine insemination. Theriogenology. 63:536-547.

Belstra, B.A. et al. (2020) Effects of the number of sperm and site of uterine semen deposition on conception rate and the number of embryos in weaned sows receiving a single fixed-time insemination J. Anim. Sci. 98(9):1-10.

Llamas-López, P.J. et al. (2019) A new device for deep cervical artificial insemination in gilts redcues the number of sperm per dose without impairing final reproductive performance. J. Anim. Sci Biotech. 10:11.
Dziuk, P.J. and Gehlbach, G.D. (1966) Induction of ovulation and fertilization in the immature gilt. J. Anim. Sci. 25:410–413.
Cassar, G. et al. (2005) Effect of single of double insemination on fertility of sows bred at an induced estrus and ovulation. J. Swine Health Prod. 13(5):254-258.
Bennett-Steward, K. et al. (2007) Ovulation induction protocol using equine chorionic gonadotropin and porcine luteinizing hormone in the weaned sow. J. Swine Health Prod.15(4):194–197.
Knox, R.V. et al. (2017A) Effects of estrus synchronization using Matrix® followed by treatment with the GnRH agonist Triptorelin to control ovulation in mature pigs. Anim. Reprod. Sci. 185:66-74.
Knox, R.V. et al. (2017B) Effects of numbers of sperm and timing of a single, post-cervical insemination on the fertility of weaned sows treated with Ovugel®. Theriogenology. 92:197-203.
Ulgium, R.R. et al. (2018) Ovulation and fertility responses for sows receiving once daily boar exposure after weaning and Ovugel® followed by a single fixed time post cervical artificial insemination. Theriogenology. 105:27-33.
Bergfeld, J. et al. (1976) Experimental biological engineering to synchronise ovulation in gilts III Onset of ovulation, ovarian findings and fertilization following methallibure, 750 IU PMS with and without 500 IU hCG at different times of treatment. Arch. Exper. Vet.-med. 30:471-480.
Brüssow, K.P. et al. (1990) The relationship between the surge of LH induced by exogenous Gn-RH and the duration of ovulation in gilts. Reprod. Domest. Anim. 25:255–260.
van Leeuwen, J.J.J. et al. (2011) Effects of altrenogest treatment before and after weaning on follicular development, farrowing rate, and litter size in sows. J. Anim. Sci. 89:2397-2406.
Fernandez, L. et al. (2005) Reproductive performance in primiparous sows after postweaning treatment with a progestagen. J Swine Health Prod. 13:28–30.
Knox, R.V. and Rodriguez-Zas, S.L. (2001) Factors influencing estrus and ovulation in weaned sows as determined by transrectal ultrasound. J. Anim. Sci.79(12):2957–2963.
Lucy, M.C. et al. (2001) Ovarian follicular growth in sows. Reprod. Suppl. 58:31–45.
Labrecque, J. and Rivest J. (2018) A real-time sow behavior analysis system to predict an optimal timing for insemination. 10th International Livestock Environment Symposium.
Klopfenstein, C. et al. (2016) Validation of the period for insemination indicated by PigWatch, relative to sow ovulation. Centre de développement du porc du Québec inc. Retrieved from http://www.cdpq.ca/getattachment/Recherche-et-developpement/Projets-de-recherche/Projet-210/Rapport-PigwatchFinalEN.pdf.aspx.
Moore, G. and Hasler, J.F. (2017). A 100-year review: Reproductive technologies in dairy science. J.Dairy Sci. 100(12):10314-10331.
Nielsen, R.L. (2021) Historical corn grain yields in the U.S. https://www.agry.purdue.edu/ext/corn/news/timeless/yieldtrends.html

Conversion to Loose Sow Housing: Polar Pork Farms Experience

Daryl Possberg

Polar Pork Farms

Polar Pork Farms is a pork production group in Saskatchewan made up of 7 farms totaling approximately 18,000 sows. There are 6 commercial farms and one multiplier. Our primary business is isowean production.

We made the decision in 2017 to convert 5 commercial operations to pen gestation. At the time one of the farms was not in production. After long discussions and planning we decided to go to a basic pen gestation system where we would put 10 sows in each pen, utilizing the existing watering and feeding system. We believed we could do this as there was a common slat adjoining the rows of gestation stalls that had a common pit. We also used the existing stalls up to the shoulder and cut the rest off. The pens were reinforced where we saw weaknesses and gates were put in to join the pens.

The first phase of the plan was to trial the system in one of our farms when we started it up. We used 1/3 of the barn when we started it up as a trial. We ran 2 gestation cycles through the trial area and decided to go implement the system in our other farms. We made some minor modifications to the program but decided it was good enough for the rest of the system.

To minimize the opportunity costs of the project, we used our empty facility to house pregnant sows while we were doing the modifications. We took animals that were 5-8 weeks confirmed pregnant and transported them to our farm not in production and continued our breeding program as normal. We would then create "construction zones" in our gestation barn where we would tarp areas off and cut holes in the wall so that renovation personnel could enter and exit efficiently. Our renovation team made offsite jigs and rented a facility to cut steel for the penning to make that process more efficient.

We were able to convert the 4 commercial operations into pen gestation without missing any services or sales commitments. The rental income from the conversions we used to help finance the start up of the 5th commercial operation. The approach of doing 5 conversions gave us scale in pricing and experience. Overall we were able to implement the conversion to pen gestation for less than $100 per sow.

The results have been positive to our production system. Sow mortality, percentage of stillborns and preweaning mortality have come down in our system, as has the time spent in maintenance for our penning. Conservatively we believe the internal rate of return on our pen gestation conversion is 26%.

The Code of Practice Five Year Review: Implications for Future Code Revisions

Yolande M. Seddon

Western College of Veterinary Medicine, 52 Campus Drive, University of Saskatchewan, Saskatoon, Canada, S7N 5B4.
yolande.seddon@usask.ca

1. The 2019 Code Review

The Code of Practice for the Care and Handling of Pigs (hereafter referred to as the Code) is the national guideline for the care of pigs in Canada and forms the basis for on-farm animal care verification programs. The Code sets Requirements and Recommendations for animal care practices, of which the Requirements are compulsory animal care standards that must be met. Adherence to the Code is verified by the animal care assessment program 'PigCARE,' within the national platform for pork quality assurance programs, Canada Pork Excellence, under which around 97% of pigs reared and marketed within Canada are covered.

The National Farm Animal Care Council (NFACC), the national lead overseeing the development of the Codes advises that per species, the Codes should be revised at least every 10 years, and reviewed every five years with amendments made as required to ensure the Codes remain current with government policy, scientific evidence and industry practice. The last revision of the Code of Practice for the Care and Handling of Pigs was 2014, therefore the Canadian Pork Council coordinated a five-year Code review in 2019. For this, a Code Technical Panel (CTP) was formed comprising representation from producers, veterinary counsel, humane societies, retail, scientific counsel in animal welfare and government; stakeholders and other expertise. The purpose of the five-year review is to: i) review the progress in the implementation of the Code; ii) review the progress in the science since the publication of the revised Code; iii) identify challenges and iv) determine the priorities for the next Code revision.

A major change in the 2014 Code was the requirement to move to group-housing for mated gestating sows and gilts. Section 1.1.2 of the Code required that all holdings newly built or rebuilt or brought into use for the first time after July 1, 2014 must house mated gilts and sows in groups, with the exception that individual breeding stalls may be used for up to 35 days from the date of the last breeding (NFACC, 2014), in order to manage the breeding and regrouping of sows whilst giving consideration to the sensitive implantation period. The Code requirement provided a deadline of July 1, 2024 for mated gilts and sows to be housed in groups or individual pens. However, given uncertainty of challenges around the transition process, if this had not been achieved, the Code permitted animals to be housed in individual stalls if they are provided with the opportunity to turn around or exercise periodically, or through other means are allowed a greater freedom of movement (NFACC, 2014). This latter requirement for periodic exercise provides a grandfather clause, permitting the continued use of existing stall barns beyond 2024 until their usable life has expired in which case the facility would need to be converted to group-housing. This approach to graduate barns to group-housing as buildings come to end of their usable life would help to manage the financial challenge that comes with building a new barn, providing time for existing loans on facilities to be paid off and the equipment invested to be used. The requirement for stall-housed animals to be provided with a greater freedom of movement directly targets a major societal concern over the close confinement of animals (Spooner et al. 2014); periodically providing a greater freedom of movement could help to address this concern. This requirement for a greater freedom of movement for stall-housed animals also applies to boars from 2024 (section 1.1.6). Suitable options for providing a greater freedom of movement were to be clarified by participating stakeholders by July 1, 2019 (NFACC, 2014), and this was a major focus of the 2019 Code review.

2. Options for Providing a Greater Freedom of Movement: Scientific Evidence

To determine suitable options to provide a greater freedom of movement to stall-housed animals required an understanding of what welfare benefits the animals could received and also what would be feasible for a producer to implement. A review of the scientific literature determined that to apply a greater freedom of movement in existing stall barns, thus not requiring building conversions, simple approaches would need to be taken, with options including removing animals from stalls and individually walking them around the barn in the alleyway between stalls, exercising animals on a treadmill, removing animals from stalls to a temporary pen, or modifying stalls to enable animals to turnaround.

Turnaround stalls were a modified gestation stall with a hinged rear portion that swings open when pushed by the sow, increasing the available floor space. The limited research that exists identified that gilts and sows housed in turnaround stalls would increase their activity, turning around on average 200 times per day irrespective of food and water location (Boe et al. 2011), would stand and nose/lick the bars of the stall more frequently, and have lower plasma cortisol levels (Bergeron et al. 1996). Despite the increased activity and potential stress reduction, turnaround stalls were not considered an ideal solution due to reports that these stalls required high maintenance, with frequent breaking of movable parts being pushed by sows, and additionally, that some sows would regularly trap adjacent animals in a smaller stall area by swinging their stall over and lying down. Intermittently penning groups of stall-housed sows together to provide freedom of movement was another option that was not deemed viable due to the risk of increased aggression between sows intermittently grouped together. Therefore, providing a greater freedom of movement, or periodic exercise to individual sows one at a time was the strategy that offered the most workable solution.

Considering providing periodic exercise in the alleyways between the stalls, it was desirable to have a measure of how highly sows and gilts valued the opportunity to be intermittently released from their stall. This could provide some information to gauge how important this requirement is for sow welfare and how frequently access to a greater freedom of movement should be given. Research was conducted to specifically ask this, with sows and gilts trained to press a button to 'ask' for time out of the stall. Sows and gilts were found to ask for time out of their stall daily by pressing the button, with gilts having a level of motivation to exit the stall equal to that of receiving a feed reward, whilst sows showed a greater motivation to access a feed reward, and a moderate level of motivation to exit their stall (Tokareva et al. 2021). Providing regular exercise to stall-housed pigs through walking animals in laps around the barn several times a week (2-7 laps around the barn, 2 – 5 time a week) has been found to offer some benefits; increasing bone strength and leg health (Schenck et al. 2008), reducing farrowing duration (Ferket and Hacker, 1985) and reducing sow restlessness (Harris et al. 2013). However, considering the range of herd sizes within Canada and the time required to exercise sows individually, it was concluded likely impractical for producers to exercise sows daily or several times a week, requiring large amounts of labour to provide sows with short periods of freedom of movement. Additionally, regular forced exercise was not necessarily right for all scenarios either, with evidence of individual differences in animals in their response to forced exercise; research reports a proportion of gilts as unwilling to walk around the barn (Schenck et al. 2008), and an increase in injury to hooves when high levels of exercise were given (25 laps around the barn, Schenck et al. 2008), which would also be influenced by the quality of flooring.

Exercising stall-housed sows at a much lower level was purposely studied (10 mins of walking once per week) and measures of welfare and productivity compared to sows in groups vs sows stall-housed throughout gestation, to determine if a much lower level of exercise that may be more achievable would benefit the sows. Ten minutes of walking once per week resulted in modest benefits to the productivity of stall-housed, older parity sows, with an increased number of liveborn piglets compared to stall-housed sows given no exercise throughout gestation, with exercised sows being comparable in performance to sows in group-housing (Tokareva, 2021). However, there was no measurable welfare benefits, with exercised stall-housed sows being most comparable to regular stall-housed sows in behaviour, and sows housed in groups showing improved indicators of welfare, displaying a reduced frequency of stereotypic behaviour in early and mid gestation, reduced sitting and increased lying (Tokareva, 2021). Altogether, the scientific evidence led the Code Technical Panel to conclude that neither walking sows around the barn at regular intervals nor housing animals in stalls that enabled them to turn around but still resulted in close confinement were ideal solutions to providing a greater freedom of movement. Neither option were equivalent to the freedom of choice, the amount of free movement and social interaction that sows can receive when housed in a group-gestation pen (Canadian Pork Council, 2019).

3. Code Technical Panel Recommendations

At the current rate, it is estimated that around 66% of the national herd will have converted to groups by 2024 (Canadian Pork Council, 2019). Therefore, rather than retain the requirement that stall-barns can continue post 2024 providing that stall-housed animals are periodically provided with a greater freedom of movement, which may not be viable or improve the welfare of sows, the decision was made to recommend a new deadline; that all barns must have converted to group housing to July 1, 2029. Providing a five-year extension was made on the basis it will afford additional time for the remaining 40% of the national herd to convert, helping to ensure the national herd transitions to groups in a more timely manner, whilst helping to prevent producers existing the industry due to the financial burden of converting, or rushing a conversion to less than optimal facilities. In the long run, this approach should result in greater gains for animal welfare: with more optimal facilities developed that can provide better group facilities for sows, and prevention of herd contraction. Where an livestock industry contracts, the production capacity that is lost will likely be absorbed elsewhere in the world, where welfare standards are often lower. Thus overall, it does not serve animal welfare in global pork production to result in contractions in an industry that is trying to provide enhanced animal welfare.

Given that animals would remain in stalls post 2024, a decision was made that the welfare of sows and gilts remaining in stalls should be improved. The CTP recommended that functional and effective enrichments (being both practical for a producer to implement and bringing welfare benefits for the sow), that show measurable welfare benefits to sows and gilts be made a requirement for animals that remain in stalls post 2024 (Canadian Pork Council, 2019). This requirement for enrichment is requested to be reviewed as part of the 2024 Code review.

As part of the CTP five year review a full review of the developments in scientific research in the priority areas identified in 2014 was conducted. As part of amendments to be considered in the 2024 Code review, the CTP recommended that the current space allowance recommendations for gilts and sows in group-housing be made requirements, and that environmental enrichment, management of pain and injury, space allowance for nursery pigs, management of gestating sows in groups and the relationship between genetic selection and animal welfare be made research priorities in order to help inform developments in the 2024 Code revision. Additionally, although not specifically covered in the Code of Practice for the Care and Handling of Pigs, transportation of swine was included as an animals welfare research priority. The results of these research priorities could have implications for revisions taking place in the 2024 Code.

The Code development process requires consensus between stakeholders on Code Requirements. Following a public comment period on the CTP recommendations, consensus was not able to be reached between committee members. Consequently, amendments to the pig Code are on hold whilst NFACC performs a review of the governance process. That consensus could not be reached is a challenge that has been rare but should not be unexpected. Making unified improvements in animal welfare within an industry is complex, especially when there is great expense, risk, retraining and structural changes required by an industry in order to meet requirements. Additionally, unlike with the dairy and egg industries, pork is not a regulated sector, and participation in the Canada Pork Excellence program is voluntary. Addressing the governance around the Code development process in order to move forward when consensus issues arise will have lasting implications for the Code development process and will improve the process in the long run.

References

Bergeron, R., Gonyou, H.W. and Eurell, T.E. (1996). Behavioral and physiological responses of Meishan, Yorkshire and crossbred gilts to conventional and turn-around gestation stalls. Canadian Journal of Animal Science. **76**, 289–297.

Boe, K.E., Cronin, G.M. and Andersen, I.L. (2011). Turning around by pregnant sows. Applied Animal Behaviour Science.**133**,164–168.

Canadian Pork Council (2019) Five-year review summary report Code of Practice for the Care and Handling of Pigs. Available at: https://www.nfacc.ca/codes-of-practice/pigs, date accessed October 2021.

Ferket, S.L. and Hacker, R.R. (1985). Effect of forced exercise during gestation on reproductive performance of sows. Canadian Journal of Animal Science. **65**, 851–859.

Harris, E.K., Berg, E.P., Berg, E.L. and Vonnahme, K.A. (2013). Effect of maternal activity during gestation on maternal behavior, fetal growth, umbilical blood flow, and farrowing characteristics in pigs. Journal of Animal Science. **91**, 734–744.

National Farm Animal Care Council. 2014. Code of Practice for the Care and Handling of Pigs. Available at: https://www.nfacc.ca/pdfs/codes/pig_code_of_practice.pdf

Schenck, E.L., McMunn, K.A., Rosenstein, D.S., Stroshine, R.L., Nielsen, B.D., Richert, B.T., Marchant-Forde, J.N. and Lay, D.C. (2008). Exercising stall-housed gestating gilts: Effects on lameness, the musculo-skeletal system, production, and behavior. Journal of Animal Science. **86**, 3166–3180.

Spooner, J.M., Schuppli, C.A. and Fraser, D. (2014) Attitudes of Canadian citizens toward farm animal welfare: A qualitative study. Livestock Science. 163, 150-153.

Tokareva, M., Brown, J.A., Woodward, A., Pajor, E.A. and Seddon, Y.M. (2021) Movement of more food? A comparison of motivation for exercise and food in stall-housed sows and gilts. Applied Animal Behaviour Science. **240**, 105348.

Tokareva, M. (2021) Motivated for movement? Exercise and the gestation environment on sow performance and welfare. Doctoral Thesis, available at: https://harvest.usask.ca/handle/10388/13465, date accessed: October 2021.

ADVANCING THE SCIENCE OF NUTRITION
World Leader in Nutrient Expertise
Focused on Our Customers
Delivering innovation to the industry
Biomin®
is now part of DSM
Hy•D®
Stand Strong for Lifetime Productivity
VevoVitall®
Improve Nursery Performance
HiPhos™
The Stable Phytase
Biomin® II & Biomin® AA
Absolute Protection
If not us, who? If not now, when?
WE MAKE IT POSSIBLE
NUTRITION · HEALTH · SUSTAINABLE LIVING
DSM
BRIGHT SCIENCE. BRIGHTER LIVING.

Applications of Next Generation Sequencing to Inform Disease Management Decisions for Livestock

Maria J. Clavijo[1,2], DVM, PhD, Robert Mugabi, DVM, MS, PhD[1] and Ganwu Li[1], MS, PhD

[1]Veterinary Diagnostic and Production Animal Medicine, Iowa State University, Ames, IA
[2]PIC®, Hendersonville, TN
Email: mclavijo@iastate.edu / maria.clavijo@genusplc.com

1. Introduction

Swine endemic systemic bacterial agents, such as Streptococcus suis, Glaesserella parasuis and Actinobacillus suis remain a global challenge for swine production by reducing productivity, impacting welfare and leading to an over-dependence on antimicrobials. In fact, recent data from the Iowa State Veterinary Diagnostic Laboratory show that the diagnosis and detection for all three agents has increased in the last couple of years (Table 1).[1] These agents are considered normal colonizers of the upper respiratory tract of pigs and both pathogenic and commensal strains can co-exist within and animal, complicating diagnosis. Control of pathogenic strains hinges on proper diagnosis, minimizing triggering or predisposing factors, effective vaccines based on the farm's strains, and strategic use medication. Next-generation sequencing provides the best resolution for strain characterization unlike conventional typing methods, and when combined with other relevant metadata can serve as a robust surveillance system to monitor these bacterial agents in flows. The objective of this presentation is to share some of the lessons learned on the use of next generation sequencing to improve the control of endemic agents in the field.

Table 1: Increase (+) or decrease (-) of the number of cases (total cases of the period) with bacterial detection and disease diagnoses over the total cases by a 5- or 10-year period.

Agent	2010- 2014	2015 - 2019	2010 - 2019
Streptococcus suis			
Detection	+ 2,112 (7,754)	+ 2,301 (17,504)	+ 4,966 (25,258)
Disease Diagnosis	+ 149 (1,701)	+ 612 (3,748)	+ 923 (54,49)
Glaesserella parasuis			
Detection	+ 718 (2,376)	+ 1,351 (13,638)	+ 3,536 (16,014)
Disease Diagnosis	+ 85 (1,038)	+ 428 (2,684)	+ 677 (3,722)
Actinobacillus suis			
Detection	+ 240 (1,069)	+ 321 (2,050)	+ 560 (3,119)
Disease Diagnosis	+ 52 (444)	+ 116 (717)	+ 169 (1,161)

2. Start off With the Right Samples

Succesfully isolating these agents from diseased pigs requires selecting untreated pigs exhibiting clinical signs characteristic of acute infection, since these clinically relevant agents/strains are rarely isolated from chronically infected or dead pigs. Pigs exhibiting respiratory distress, central nervous systems signs, arthritis and fever should be selected. All of these agents are commensal organisms of the upper respiratory tract (i.e. nasal cavity, tonsil and trachea), thus, non-respiratory, systemic tissues are prefered for successful culture of the relevant strain. While *S. suis* and *G. parasuis* can cause pneumonia and be successfully cultured from the lung, sometimes strains isolated from this site may not represent the disease-causing strain. Thus, brain, csf, joint tissue/fluid, serosal tissue, fibrin and systemic organs, such as spleen, liver and thoracic lymph nodes should be collected, mainly in 3-10 week old pigs. Specifically for *A. suis,* in addition to systemic organs and serosal tissues, lungs showing classical petechial to ecchymotic hemorrhages should be collected from affected pigs.

From systemic cases, submission of individually bagged fresh tissues and synovial or fibrinous fluids provides the highest likelihood of isolation of the agents. Swabs can also be collected from organs with fibrinous exudate on the surface, but some of these agents (i.e. *G. parasuis*) require the addition of Stuart or Amies media to maintain viability of the agent. Specifically for brain samples, either sending the intact head or aseptically collected meningeal swabs minimizes the risk for contamination of the sample. Submitting fixed tissue and requesting histopathology is important to align lesions with isolation or detection of the agent. Finally, it is critical to refrigerate the samples as soon as they are collected and shipped promptly to the veterinary diagnostic laboratory. Isolation of the agents within lesions in pure culture provides assurance of successful sampling collection and identification of the likely culprit.

3. Uses of Next Generation Sequencing:

- Improving how populations of pigs are mixed: The information generated can help improve how populations of pigs are mixed. The in-depth characterization of endemic agents from multiplier farms, has allowed swine practitioners to determine the most appropriate source for their replacement gilts. Within production systems, multiple-source commingling is common practice. NGS analysis can help determine which flows are more compatible for mixing based on their profile of endemic agents.

- Autogenous vaccines: Vaccine candidate selection and efficacy for agents such as, *S. suis* and *G. parasuis* has proven to be very innacurate in the past. Efficacy hinges on having the right antigen, the right antigenic mass and a robust adjuvant. While more validation and experimentation is needed to confirm bacterin efficacy for these agents, based on recent feedback from swine practitioners, when the right strain is selected these products appear to be efficacious. Use of NGS in many field cases has revealed that the *S. suis* strains included in the current autogenous vaccine did not match the clinical isolates obtained from contemporary cases, evidencing that improvements in the vaccine selection process were warranted.
- Improving how we use antibiotics: We are still scratching the surface on the application of the antimicrobial data obtained with these platforms. However, AMR data obtained and correlated with clinical and antimicrobial susceptibility data can help identify the most suitable antimicrobial for on farm use, and help track changes in prevalence of resistant strains. So far, the distribution of AMR genes has helped identify the clinically relevant strains.
- Surveillance: The use of NGS can allow for vast improvements in on-going surveillance programs for these agents. For example, a baseline of current strains in a flow can be obtained and routine incorporation of clinical strains can help identify the emergence of new pathotypes and changes in prevalence, as well as support the elimination of disease-associated strains from pig populations.

4. Lessons Learned - *S. suis*

A total of 531 *S. suis* isolates submitted to the ISU-VDL were whole-genome sequenced and analyzed. These were submissions from clients who submitted isolates (2014-2020) from disease associated cases. A total of 25 different serotypes were identified, 1 (n=78), 2 (n=45), 9 (n=32), 5 (n=27), 23 (n=46), 1/2 (n=24), 8 (n=19), 7 (n=18), 14 (n=16), 4 (n=14), 3 (n=11). The rest of the serotypes 10, 11, 12, 13, 15, 16, 18, 19, 21, 24, 28, 29, 30 and 31 were represented by 1 to 8 isolates (Figure 1). A total of 183 isolates were untypeable representing 34.5% of the isolates (Figure 1). Multi-locus sequence typing (MLST) identified 36 sequence types (STs) with the predominant being ST1 (n=140), ST28 (n=70), ST108 (n=30), ST977 (n=22), ST25 (n=12), ST87 (n=14), ST94 (n=13), and ST961 (n=12) (Figure 1). Although certain STs were more associated with specific serotypes, for example serotype 1 (n=78) and serotype 14 isolates were almost exclusively ST1, ST diversity was noticed in 7/25 serotypes with the number of STs ranging from 2-6 per serotype. These serotypes included serotype 2 and 1/2, 3, 4, 5, 7, 9 and 23. But within these diverse serotypes, a given ST usually predominated (26/45 serotype 2 isolates were ST28). It is important to highlight that ***serotyping can tell an incomplete picture on the clinical relevance of an isolate.*** For example, in our database, we identified three distinct genotypes of serotype 2, that given other genetic characteristics (i.e. ST, distribution of VFs and AMR genes) can either be characterized as high, low or medium pathogenicity. Thus, further characterizing these strains is a critical step for the identification and management of clinically important isolates within a flow. Finally, a total of 147 (27.7%) of the isolates could not be typed by MLST, and likely represent novel STs. Within our database roughly 10% of isolates are untypeable by both serotyping (conventional and WGS) and MLST. The majority of these isolates either originated from polymicrobial cases (i.e. viral and bacterial agents) or the isolation of *S. suis* did not align with lesions present in the tissue, suggesting possible contamination. This indicates that these strains potentially represent opportunistic or commensal strains. Thus, it is imperative that selected pigs for submission are exhibiting acute clinical disease and have not been treated with antibiotics. From these, we recommend focusing on affected systemic tissue (i.e. brain, csf, joint swabs/tissue or serosal surfaces). It is also important to avoid using captive bolt or blunt force trauma, since those methods can increase the likelihood of cross-contamination.

Strains that are untypeable by both methods tend to additionally have two unique genetic features: ***1) Overabundance of antimicrobial resistance genes (AMR)*** and ***2) a lack of putative virulence factors,*** specifically they lack the traditional *mrp*, *epf*, and *sly* virulence-associated genes. Still, the clinical relevance of these isolates is yet to be determined. In contrast, majority of strains with a defined serotype and ST usually carried at least 1 of the traditional virulence markers (i.e. *mrp*), harbored a significant amount of putative virulence factors and had significantly less AMR genes. The low levels of antimicrobial resistance noted at the genomic level are in alignment with the low levels of phenotypic resistance observed at the ISU VDL. Currently, distribution of AMR genes aids in determining the clinical relevance of strains. While the data show a wide diversity of *S. suis* strains present in NA swine herds, usually one or two strains predominate within a farm or flow in disease associated cases, and ***these same strains tend to persist***

over time. Finally, in our experience, it is important that ***repeated tissue submissions be carried*** out to increase the likelihood of capturing the clinically relevant strains and have a good understanding of the diversity and frequency of disease associated strains found in a farm.

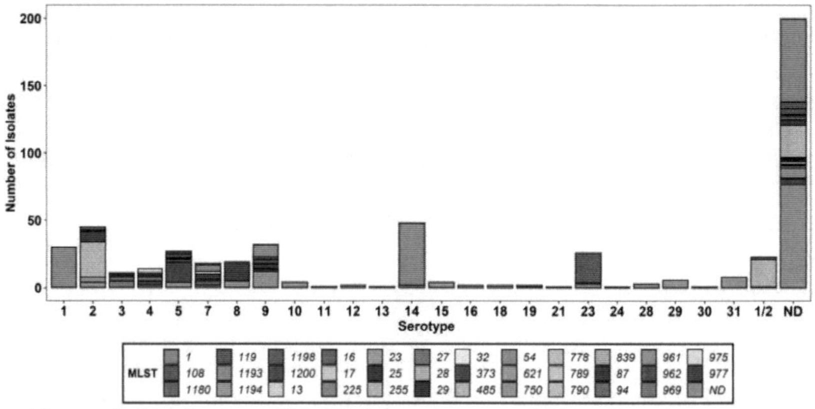

Figure 1. Distribution of MLST sequence type (ST) by *S. suis* serotype.

5. Lessons Learned - *G. parasuis*

Over 100 *GPS* isolates submitted to the VDL between 2015 and 2020 have been whole genome sequenced using Illumina Miseq platform and analyzed using an in-house bioinformatic pipeline. The most common serotypes identified were 7, 13, 4, 2, 5/12 and 1, with a clear overrepresentation of serotype 7 strains, previously thought to be avirulent. Similar to *S. suis*, multiple serotypes (i.e. 7, 4 and 2) were distributed in multiple clusters, suggesting that further genetic characterization is needed for proper identification and determining their relevance. MLST revealed 45 different STs including ST454 (n=18), ST6 (n=8), and ST415 (n=6) and the rest with 1-3 isolates each. Interestingly, most strains in the database were assigned novel STs by pubmlst, highlighting potential genetic differences between strains from Europe, Asia and North America. In terms of virulence-associated markers, the majority of isolates from serotype 7, 3, 4, 13, 5/12, carried at least 5/9 of group 1 vtaA genes (trimeric autotransporters) and most carried group 3 vtaA genes. While serotype 6 is typically regarded as non-pathogenic, this serotype was detected in multiple pigs from one flow experiencing *GPS*-associated disease and carried most of vtaA group 1 genes, suggesting that not all serotype 6 strains are non-pathogenic. In contrast, only 3 isolates in the database (serotypes 8 and 15) did no carry any of the vtaA genes and clustered with reference strains from serotype 6, 10, 3, 9 and 8, previously described as non-pathogenic. In terms of AMR genes, most isolates were positive for *bcr*, *aph(3")-Ib*, *bacA*, *ksgA*, and *sul2* genes which are markers for bicylomycin, streptomycin, bacitracin, kasugamycin, and sulfonamides, respectively.[2]

6. Conclusions

Endemic agents continue to challenge the health and productivity of swine production systems and, thus, there is a need for effective and implementable solutions to address these challenges. The creation of a large, well-characterized database of isolates from swine herds provides valuable and actionable information for gilt sourcing, vaccination and antimicrobial therapy. While next generation sequencing can provide significant value for control programs, it is important to highlight that the success of managing endemic agents requires a multifaceted approach, that includes, comprehensive veterinary clinical work-up, minimizing triggers, such as co-infections, selecting the correct pig and sending high quality samples, and continuous monitoring of pig populations.

References

Silva et al., 2021. Detection and diagnostic trends of five swine endemic bacterial pathogens (2010-2019). *AASV Proc.* 2021.

Mugabi et al., 2021. Characterization of *Glaesserella parasuis* strains circulating in US swine herds using whole genome sequencing. *AASV Proc.* 2021.

Poop, Blood, Spits, and Others: the Art of Sampling Pigs

Karine Talbot

HyLife

There are many reasons why we need to collect samples from pigs, and there are many different techniques and type of samples that can be collected. Over the years, some of these techniques have become more practical and easier to perform on-farm, by almost anyone. But to be useful to your veterinarian, the right type of samples for the right pigs, collected for the right reasons must be done... It is not as hard as it sounds, but it definitely needs to be done correctly!

Why do we need to sample pigs? There are many reasons, but the most common are the following:

- To prove negative status or for early detection: this is mostly used in high health herds or multiplier herds (for example PRRS and Mycoplasma hyopneumonia negative) to ensure they stay negative over time; or in an isolation barn, to confirm the negative status of gilts before moving them to the sow barn, for example. When testing to prove negative, the assumption is that the disease we are testing for is not present. It usually requires more samples and techniques that are more sensitive.
- Disease investigation: this is usually done when pigs are showing clinical signs and samples need to be collected to determine what disease is present. The type of samples collected in these cases usually depends on

the type of clinical signs observed (respiratory (ex: coughing), digestive (ex: diarrhea), systemic (ex: sudden death)).

- Disease monitoring: this is usually done to assess the status of a disease in a herd that is known positive. For example, it could be to assess the PRRS circulation in piglets weaned from a sow barn vaccinated to PRRS; or regularly collecting samples from a herd to assess the circulation of influenza and possibly make autogenous vaccine from isolates found in the herd.
- Eradication: in this case, samples are collected to monitor the progress of eradication protocols, including before, during and after the eradication.
- Vaccine compliance and efficacy: samples can also be collected to assess the seroconversion of pigs to certain vaccines and understand their vaccination status.

Once we have determined why we need to sample pigs and what type of diseases we are looking for, the next step is to understand and choose the test that will be required. There are various types of tests offered by diagnostic labs, and they all have their uses. The most common are the following:

- Serology: serology tests are very common and are offered for a wide variety of pathogens. Serology test measures the level of antibodies in the serum, which indicates if the pigs have been exposed, or vaccinated, to a specific disease. They do not differentiate between a natural infection or a vaccination. Serology is mainly done on serum (bloods), and some other types of samples for specific pathogens (example: PRRS, which can now be done on oral fluids in some labs; PCV2 on milk or colostrum, etc…). Serology results can sometimes be difficult to interpret, and false positive are common.

- PCR (antigens): Polymerase Chain Reaction (PCR) detects the antigen (the pathogen) in the samples. PCR exists for many diseases and are very sensitive (detect the pathogen presence at very low level). A variety of samples can be used to test with PCR; tissues, serum, fecal samples, oral fluids, processing fluids, environmental samples, etc… It is also a very specific test with rare false positive. However, the PCR cannot tell you if the pathogen present is actually infectious (alive); a positive result means that the genetic material of the pathogen was found, but it could be dead and not infectious (something important to know when using PCR for environmental surveillance).
- Culture (bacterial/viral): Laboratories can do culture for bacterial diseases or some viral diseases. A culture is when the sample is put in a special environment (ex: culture plate or broth) in a lab to encourage growth of certain cells. If certain cells grow, they then need to be properly identified by the lab to confirm what pathogen it is (ex: E.coli, salmonella, Strep.suis, etc…). Culture is also required in some cases if the objective of sampling is for autogenous vaccine purposes.
- Histology: Histology is when a fixed tissue is observed under a microscope by a pathologist to assess the presence or absence of lesions. Some lesions are indicative of specific diseases, and some are more general. Histology requires tissue samples.

Once we know why we are testing, what we are testing for and how we will test for it, it is important to select the correct type of samples. Some examples of samples are; serum (blood), oral fluids (saliva), feces, tissues, environmental samples, etc…

Not every sample type is appropriate for all diagnostic tests, or for any diseases. For example, influenza will not be found in a fecal sample… and the lab cannot perform histology on a blood sample! You might find circovirus in an oral fluid sample, but it might not mean that it is causing an issue in your pigs… so collecting the right type of samples is critical to properly answer the diagnostic question you have!

Once you and your vet have determined what you need, it is time to collect the samples! Here are various techniques that can be used on-farm, with their pros and cons.

Necropsy (tissues): Necropsies are usually performed by a veterinarian, a technician or someone that has been trained to perform them. It requires a higher level of training than other sampling techniques. Usually, necropsies are performed only on a few pigs that are dead or euthanized, so the sample size is small. It is most appropriate for disease investigation, when we need to see lesions in the pigs, or when various tests are needed on various types of tissue samples (PCR, bacteriology, histology). Necropsies (and proper tissue collection) are very useful for the diagnostic of any diseases. Alternatively, a whole pig carcass can be brought to a diagnostic lab for the pathologist to perform the necropsy and collect tissues.

Blood (serum): Blood samples are usually collected by a veterinarian, a technician or someone that has been trained to perform them. It requires a higher level of training than some other sampling techniques but a lot of people, when given the opportunity, can learn to collect blood samples quite well. Blood samples can be collected from a large number of animals, but the pigs must be restrained for the sampling (either snared or hold, depending on the size of the pig), so it usually requires 2 people. Blood samples are the main type of samples that can be used for serology. Bloods are useful to confirm a negative status, do surveillance or early detection, monitor some diseases in the herd, for disease investigation, throughout an eradication program or for vaccine compliance.

Oral fluids (saliva): Oral fluids can be performed by anyone, as they are very simple to collect. They consist in hanging a cotton rope into a pen of pigs, let them chew on it for 15-20 minutes, and then wring the saliva out of it to send it to the lab. There is no need to catch the pigs one by one like for blood collection, and a large number of pigs can easily be sampled on each rope. Oral fluids are mainly used for PCR tests (a few labs are now offering PRRS serology), and several diseases can be tested on it (PRRS, influenza, PEDV, etc…). Oral fluids are used to confirm a negative status, do surveillance or early detection, monitor some diseases in the herd, for disease investigation and throughout an eradication program. However, if a positive sample is obtained, they are not always the best samples to perform further test like sequencing or isolation as they are heavily contaminated (fecal material, feed, dust, etc…) and the sample degrade quickly.

Wipes (nasal, udder): Wipes are easy to perform and can be done by anyone. Nasal wipes are simply done by using a wipe (usually a gauze; must be the right type of wipes though) and passing it over the noses of the pigs. Udder wipes use the same process but passing it on the udder of the sow instead. The same wipes can be used for a couple pigs and many animals can quickly be sampled by one person. The wipes are then placed into a solution (sterile water, PBS) and sent to the lab for PCR testing. They are mainly use for influenza testing but can work for coronavirus like PEDV as well.

Fecal swabs and bootie: Fecal swabs or over-the-shoe booties are very easy to perform by anyone; the swab consist of a Dacron swab dipped into fecal material found on the floor or in the rectum of individual pigs; the booties are a pair of booties worn to walk from pen to pen and squeezed in sterile water after. This technique is great to sample a large number of pigs very quickly and is very sensitive to detect coronavirus like PEDV. The fecal swabs can also be used for bacterial culture. These 2 types of samples are however very specific to the detection of enteric diseases.

Processing fluids: Processing fluids are easily collected by placing all testicles and tails from processing into a bag and collecting the liquid cumulating at the bottom of the bag into a tube to send for PCR testing. It is very easy to perform and is great to sample a large number of piglets without extra effort. Processing fluids are useful to confirm a negative status, do surveillance or early detection, monitor some diseases in the herd, and throughout an eradication program. They are, however, very age specific (3-7 days old).

Tracheal or Laryngeal swabs: Tracheal and laryngeal swabs are usually performed by a veterinarian, a technician or someone that has been trained to perform them. They are not very difficult to do and can be taught easily. Like blood collection, the pig sampled must be either snared or hold, so it requires 2 people. These types of samples are great for mycoplasma hyopneumonia diagnostic on live pigs and are more sensitive with less false positive or false negative than other sampling techniques. These types of samples are also excellent for influenza diagnostic and isolation.

Others: Listed above are the most common sampling techniques used in swine. There are many others less commonly used; for example, colostrum or milk samples, semen samples (common in boar studs), blood drop, ear punches, etc...

There are multiple types of samples and techniques that are available to swine producers and their veterinarians. However, not all samples are fitted for all types of tests or to diagnose all diseases. The key to successful diagnostic is to know why you are testing, what disease you are testing for (or at least have a general idea like respiratory vs enteric) and know which test needs to be performed (serology, PCR, etc…). Then samples must be collected properly and handle properly (keep them cool and protected from sunlight!). With these few recommendations in mind, you will obtain valuable diagnostic and will be more confident that your negative results are truly negative or that you will find what you are looking for!

Where to find more information on various sampling techniques:

Oral fluids:

https://www.cfsph.iastate.edu/pdf/oral-fluid-collection-in-pigs

https://www.securepork.org/Resources/Oral-Fluid-Collection-Booklet.pdf

https://www.youtube.com/watch?v=u-uZLcZLFIl

https://vmdl.missouri.edu/testing-oral-fluids/

Tracheal swabs, laryngeal swabs:

https://www.youtube.com/watch?v=RzL9_G3naAs

https://vetmed.iastate.edu/story/vdl-tracheal-sampling

https://vetmed.umn.edu/centers-programs/swine-program/research-sdec/sample-type-detection-mycoplasma-hyopneumoniae/oro-pharyngeal-swabs

Vaccine Program and On Farm Application: How to Get the Most Out of Your Investment

Dr. Cordell Young

Precision Veterinary Services
Red Deer, AB
Email: cordell@precisionvet.ca

1. Introduction

For the most part, vaccine protocols and vaccinations are not a new concern to the modern swine producer in North America. Vaccines continue to develop and offer more protection to a wider scope of diseases. The main drivers for vaccination development and usage on farm is to view animal health in a preventative way. Vaccines are a form of investment and insurance policy to swine producers in a time when profits are squeezed and the margin for error and need for precision farming are required to continue operating. Additionally, vaccines and preventative medicine of any kind is important to help swine operations and the swine industry reduce antimicrobial usage. To accomplish this, all measures need to be taken to allow vaccines to gain full protection in the animals at the time they face disease challenges. Of course, most of the suggestions need given in the presentation need to be confirmed, validated, and approved by your herd veterinarian.

2. Shipping, Receiving, and Storage

Typically, a vaccine intended for refrigeration storage will be rendered useless if frozen. These should be reviewed for accuracy often in the barn. The labels will most commonly read "*Store in the dark at 2 to 7-8 degrees C (35-45F). Do not freeze.*" or a variation on that theme. Other vaccines will require staying frozen at -80C in a specialized freezer or yet others -20C in a regular chest freezer. In each case, it is important to make sure the storage is correct. This is important to be validated at the farm level; not just that the fridge or freezer is working, but that the temperature is correct in the precise location the vaccines are being stored. Thermometers and data-loggers are inexpensive and readily available to support producers. Producers may see value in running a log to ensure refrigerators and coolers are at desired temperatures on a regular basis.

When receiving product from your veterinarian, depending on the path it takes and shipping routes, data-loggers may be used to confirm the proper temperature during transport. If these are not accessible, a practical measure for temperatures when receiving product to the farm is an IR thermometer gun. In general, freezing is more detrimental to product efficacy than becoming too warm, but both scenarios reduce confidence that the product will work and certainly shelf life.

3. Administration

An old saying from the nutrition world is "There are 3 rations; the ration formulated by the Nutritionist, the ration mixed and provided to the animals, and finally the ration that the animals ingests." This concept correlates to swine producers and veterinarians; the vaccination protocol, the vaccine that is injected into the pig, and finally the level of protection that faces the disease challenge in the population all need to be reconciled. The obvious goal is for each step to set the animals up with the most protection possible from a given vaccine measure. In general, we expect vaccines to reduce the severity of disease, degree of shedding, and possibly duration of illness for some diseases. Preparing the vaccines before administration may take several hours up to a day, depending on the vaccine. For example, vaccines administered through the water should have the stock solution calculated at a similar time the day before administration. For injectable vaccines, slowly warming the product in lukewarm water prior to administration is important to reduce the reaction rate in the animals injected. With approval from your veterinarian, animals with severe reactions to vaccine may recover with administration of a steroid or epinephrine. However, these options should be as a last resort due to the systemic effects on the pig.

Food safety and safety for the handler is very important during the administration of any product via needles. Needle sizes should comply with the CPC Guidelines in the CQA/CPE Producer Manuals. Ensure that staff are trained, animals are restrained properly, and that the process is done at a speed that staff are not subjected to the risk for needlestick injuries. In the event of such an injury, staff need to be encouraged to contact a healthcare provider. Having staff that are up to date on Tetanus vaccinations and First Aid Kits close at hand will be an asset in the event of staff injuries.

4. Determination of Need

Acknowledging the costs of vaccine and labor, as well as the risk and stressors involved with the vaccination process, adding a vaccine to the protocol requires a sufficient objective. For some diseases, the producer and veterinarian will observe clinical signs, lesions, abnormal behavior, reduction/loss of production or a slaughter check performed at the plant. In some cases, with sub-clinical disease, producers and veterinarians may feel that implementing the vaccine, considering cost and other factors, still provides return on investment and peace of mind knowing that preventative measures were taken should a flare-up or irregular challenge present itself. A valuable tool to evaluate need, and later confirm success of a vaccine is through diagnostics like serology, PCRs, cultures, and sequences.

In addition to choosing which disease to vaccinate for and which product, timing is also pertinent. Timing of vaccination needs to consider the time it takes until the animal has enough protection to face a challenge, other stressors that are present in the animal at the same time, the protection that may come from their mothers, and duration of immunity. In general, the more often an animal "sees" an antigen, the more robust protection they will develop. To accomplish this, many products have a second dose on the label as a booster. Others, such as some circovirus vaccines, have the capability of splitting the vaccination such that between two injections.

5. Optimizing Vaccine Capability

Unfortunately, it is a reality that no vaccine will work 100%, providing 100% immunity or protection, and for an unlimited duration of time. However, we want to do everything we can to optimize the response to vaccination to get the most out of our investment. If the pigs we are vaccinating undergo many other stressors at the same time, their ability to build the immunity needed will be reduced and the potential for vaccine failure or "break-through" increases. Internal biosecurity contributes significantly to the amount of disease load that exists in a barn. Good internal biosecurity includes a long list of management practices, such as washing and disinfecting rooms between batches of pigs, boot washing, and changing clothes between compartments. A significant threat to producers is a syringe that is used week after week with no, or inadequate, cleaning. Instituting rodent control programs also has significant potential to reduce pathogen load. For some diseases, ensuring protection in the sow herd is robust will set the growing herd up for lower challenge and subsequently less disease. In barns where pig flow requires more moves and moves through multiple areas of the barn, the stress and disease exposure will impact vaccine efficacy. Pig flows that are continuous flow have additional challenges that all-in-all-out flows do not. A challenging time in a pig's life is weaning, which, for convenience is often chosen for vaccination. If the disease burden at this time is high and immunosuppression is present, the ability for the pig to develop immunity from the vaccine will be hindered.

In summary, we must minimize the number of stressors, disease and otherwise, through on-farm management practices that can be implemented as a foundation for an effective vaccination program.

References

Hess, A. (2019). Practice Needle Know-How. National Hog Farmer.
Husa, J. (2002). Avoiding Vaccine Handling Mistakes.
Canadian Pork Council. (2018). CPE Manual.
Tully, B. and B. Chappell. (2019). Internal syringe cleanliness trials.

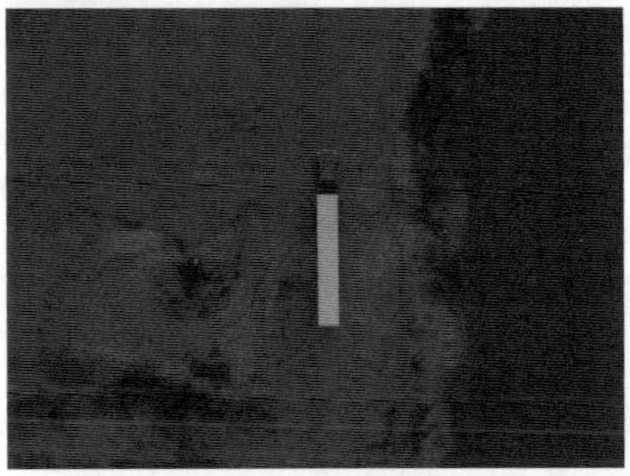

STEVE'S

ABSTRACTS: Best of Current Research

Microbiology and Immunology

Nutrition

Pig fecal microbiota during early stages of production

Maddison Arsenault[1], Brandon Lillie[1], Khurram Nadeem[3], Ehsan Khafipour[4] and Vahab Farzan[1,2]

[1]Department of Pathobiology; [2]Department of Population Medicine; [3]Department of Mathematics and Statistics, University of Guelph, Guelph ON N1G 2W1;
[4]Cargill Health Technologies, Cargill Incorporated, Wayzata, MN 55391
Email: marsen04@uoguelph.ca

Microorganisms that colonize swine gastrointestinal tracts form a complex community referred to as the microbiota. Targeting microbiota by promoting development of beneficial bacteria and communities is a potentially powerful tool for improving swine performance and health. The objective of this study was to assess composition and progression of piglet fecal microbiota from birth until weaning. Nine cohorts of 40 pigs (n = 360) from distinct farrowing sources in Ontario and Quebec were included in this study. Five cohorts were raised conventionally and four cohorts were raised without antibiotics. Fecal samples were collected within the first week of birth and continued weekly until one-week post-weaning. Samples were subjected to DNA extraction and sequencing of the V3 to V4 regions of 16S rRNA genes to identify bacterial features. Fecal microbiota was dominated by Firmicutes (41.8 %), Bacteroidetes (32.4 %), and Proteobacteria (14.1 %). Three successive stages of microbiota were identified including stage one (1 to 7 days old), stage two (8 days old to weaning), and stage three (weaning to one-week post-weaning). Each stage was associated with a distinct community cluster. Data suggest that fecal microbiota undergo several shifts throughout early stages of production. Further analysis will examine associations between piglet microbiota and antibiotic use, sow microbiota, and pig genotype. These findings will allow identification of external variables that can be used to manipulate microbiota variation and improve health.

Implications: Examining microbiota progression at early production stages is the first step in creating opportunities to target microbiota composition and maximize health and productivity. These findings will help implement interventions towards promoting health and performance of pigs, decreasing antimicrobial use, increasing animal welfare, and having more profitable swine production.

Disease-causing *Brachyspira hampsonii* efficiently harvests iron, likely outcompeting beneficial bacteria

Hannah Burlet and Matheus Costa

University of Saskatchewan
hannah.burlet@usask.ca

Bloody diarrhea and colitis in pigs, caused by pathogenic *Brachyspira hampsonii*, prevents optimal animal growth and profit as well as impactsanimal welfare. Iron is one of the main limiting nutrients for *Brachyspira* growth, and thus it could be related to disease. We have identified two proteins in the *B. hampsonii* genome thought to be associated with collecting iron: 3A and 3B (P3A/B). We hypothesized that pathogenic strains express more of these proteins than non-pathogenic strains, leading to more efficient growth disease. Using blood as a source of iron, *Brachyspira* were cultured in broth (brain-heart infusion, BHI) without blood (BHI), with blood (BHI+blood), or with blood added after 24 hours of culture (BHI+blood 24h), with 3 replicates per treatment. Cultures were incubated at 39ºC for a total of 72 hours, then total RNA was extracted and transcribed into cDNA. Gene expression levels for Proteins 3A and 3B were investigated using a novel reverse-transcriptase real-time PCR (RT-qPCR) protocol. Statistical analysis was done using the MCMC.qPCR R package. It was determined that the pathogenic strain expressed significantly lower levels of P3A than the non-pathogenic strain, regardless of the treatment ($P<0.05$). In contrast, the pathogenic strain expressed significantly higher levels of P3B than the non-pathogenic strain when there was no blood added to the media (BHI, $P<0.05$) or when blood was added at the beginning of the incubation period (BHI+blood, $P<0.05$). When blood was added after 24 hours, Protein 3B was downregulated significantly ($P<0.05$) in the pathogenic strain compared to the non-pathogenic strain. It can be concluded that pathogenic *B. hampsonii* efficiently produces iron-harvesting proteins appropriate to iron levels in the environment, making it more adaptable than its non-pathogenic counterpart. The pig and the microbiota compete for the very little free iron on the intestine.

Implications: Microbiota modulation to support this competitive environment is suggested as an approach to mitigate *Brachyspira*-associated disease. Specifics regarding this modulation remain to be unveiled but could revolve around dietary or probiotic supplementation.

Ear-tip necrosis is not associated with virus, likely due to bacterial infection

Matheus O. Costa[1,2*], John C. S. Harding and Roman Nosach

[1]Western College of Veterinary Medicine, University of Saskatchewan, Saskatoon, SK.
[2]Faculty of Veterinary Medicine, Utrecht University, The Netherlands.
[*]matheus.costa@usask.ca

Swine ear-tip necrosis (ETN) is a disease of global presence and unclear aetiology. Little evidence is available regarding the nature of this disease. The aim of this work was to investigate if ETN is an infectious disease that could be replicated using a lesion macerate inoculum. A source farm with a history of ear-tip necrosis was identified and five weeks-old pigs (n = 12) from this farm were housed under controlled conditions and intradermally inoculated with ETN lesion macerates (right ear, n = 10) or sterile inoculum (left ear, n = 10). Two pigs were not inoculated, serving as sentinels. All animals were clinically monitored daily for 21 days, and a ETN ear score was used to follow disease progression. Anaerobic (n = 2) and aerobic (n = 2) overnight cultures, as well as raw aliquots of the lesion macerate inoculum (n = 2) and control inoculum (n = 2) were submitted for metagenomic sequencing. All inoculated ears developed lesions suggestive of early ETN, but none progressed to result in loss of the ear pinna. All completely resolved 21 days post-inoculation. Metagenomic analysis identified the presence of previously suggested bacterial etiological agents, but no relevant viral, fungal or protozoan agents in the inoculum. ETN etiology remains unclear, but an infectious cause and bacterial agents are suggested to be at least partially implicated in disease pathogenesis. Virus and fungi do not seem to significantly contribute to disease.

Implications:

Artificial intelligence for Enzootic Pneumonia and Atrophic Rhinitis lesions evaluation

Mike DeGroot[1], Ramon Jorda[2] and Irene Galé[2]

[1]HIPRA Canadá, K1Y4S1 Ottawa (Canadá) [2]HIPRA HQ, 17170 Amer, (Spain).
Email: mike.degroot@hipra.com

The assessment of pulmonary lesions compatible with Enzootic Pneumonia and Atrophic Rhinitis (AR) at the slaughterhouse is one of the techniques for assessing the incidence of these diseases on pig farms and helps assess the effectiveness of vaccines against *Mycoplasma hyopneumonia* (*Mhyo*) and *Bordetella bronchiseptica* (Bb) and *Pasteurella multocida* type D (PmD). Until now, monitoring the degree of pulmonary involvement caused by *Mhyo* and snouts lesions required highly skilled technicians at the slaughterhouses. The process is undoubtedly slow and time-consuming and the disparity of the pulmonary assessments carried out by different technicians reveals this is a subjective process. Artificial intelligence (AI) is defined as the ability of a machine to present the same capacities as human beings. Artificial Intelligence Diagnos (AI Diagnos) is a fully automated diagnostic system, which allows to simplify, objectify, and facilitate the whole process of slaughterhouse assessment of *Mhyo*-compatible lesions in lungs and of Atrophic Rhinitis in nasal turbinates. The detection of objects is a smart image recognition technique which allows to identify and locate objects within an image. AI DIAGNOS has two detectors per model: the detector of focus, which identifies the lung or the nasal septum, and the detector of areas of interest, which identifies the various areas of the lung or the nasal septum. The model takes each and every image for analysis and shows the lesion grade for each section. The system has been trained using over 11,000 images, corrected for learning by experts in Mhyo and AR, and validated. The result of this validation was that the system currently has more than 85% direct accuracy by system-evaluator and more than 96% with an error per lobe rate of ±1.

Implications: Artificial Intelligence Diagnos is a great breakthrough for the swine industry, providing a new system to evaluate lesions caused by Enzootic Pneumonia and Atrophic Rhinitis with has very clear advantages: automated, reliable and simple.

Bordetella bronchiseptica, new detection techniques for a re-emergent bacterial pathogen in an antibiotic reduction scenario

Mike DeGroot[1] and Irene Galé[2]

[1]HIPRA Canadá, K1Y4S1 Ottawa (Canadá), [2]HIPRA HQ, 17170 Amer, (Spain). Email: mike.degroot@hipra.com

The Canadian swine industry remains focused on disease prevention and antibiotic reduction. With less mass use of antibiotics, disease prevention by vaccination will become more important to prevent outbreaks of bacterial diseases such as Non-Progressive Atrophic Rhinitis (NPAR). NPAR is a bacterial disease present in all swine producing countries and is caused by toxigenic *Bordetella bronchiseptica* (Bb). Clinical signs of sneezing, ocular discharge and nasal discharge are found in young piglets (6-7 weeks old). In addition, Bb is also a primary cause of Progressive Atrophic Rhinitis due to its coinfection with Pasteurella multocida type D (PmD) and also bronchopneumonia in older pigs, contributing to the Porcine Respiratory Disease Complex (PRDC). Determining the prevalence of Bb on Canadian pig farms will be important to monitor the re-emergence of this pathogen. It has been shown that the detection rate of Bb was similar in oral fluids and nasal swabs from the same animal or group of animals. In addition, the feasibility of using dried porcine oral fluid for the diagnosis of Porcine Atrophic Rhinitis has been described. Using this technique of oral fluids for PCR testing of Bb will provide a quick and easy method of sample collection for the surveillance of Bb in Canada. Furthermore, the use of FTA cards in sample collection allows for samples to be collected and transported easily without the need for refrigeration. Using PCR testing of oral fluids for detection of Bb has been completed on two Canadian farms at this time. Although the data set is too small at this time to draw any conclusions, testing has shown the presence of Bb on Canadian farms detected by PCR testing of oral fluids. Further testing of Canadian farms will need to be done to determine the prevalence of this pathogen. Currently there are no vaccines registered in Canada to control Bb. Rhiniseng® vaccine against Bb and PmD has been registered in several countries in Europe, Asia and Latin America and has been shown to reduce mortality due to rhinitis on farms with a positive diagnosis of Bb. A farm trial conducted in Canada also showed a reduction in nursery pig mortality after implementation of vaccine against Bb.

Implications: Since Bb is expected to be a re-emergent pathogen with a reduced antibiotic scenario, it will be important to monitor the importance of this pathogen in Canada. Having the tools and resources to easily detect and prevent Bb in Canada will help reduce the production losses due to rhinitis.

Coinfections between virulence factors of *Escherichia coli* and *Clostridium perfringens* type C in Canada

Mike DeGroot, Oriol Boix and Daniel Angelats

HIPRA, Amer (Girona), Spain.
Emailoriol.boix@hipra.com

Enterotoxigenic *E.coli* (ETEC) and *Clostridium perfringens* type C (CpC) are important causes of neonatal diarrhea in piglets. ETEC produces several virulence factors such as fimbrial adhesins and enterotoxins, which are responsible for the adhesion of the bacteria in the endothelium and release of toxins in the intestinal lumen. Through vaccination, colostrum immunoglobulins block the adhesin function and bacterial attachment to the intestinal epithelium and intestinal colonization are prevented. The aim of this study was to evaluate the prevalence of ETEC and *Clostridium perfringens* type C (CpC) in Canada, as well as the number of coinfections between the virulence factors in piglets suffering from neonatal diarrhea. A total of 81 fecal samples from Canada were analyzed and included in this study. The samples had been obtained between March 2015 and October 2021 and were subsequently sent to the DIAGNOS® Laboratory using FTA® ELUTE cards (Whatman Inc., Florham Park, NJ). A multiplex polymerase chain reaction (PCR) test, adapted from previous studies, was performed to detect genes encoding F4, F5, F6 adhesion factors, the heat-labile toxin (LT) of ETEC and β-toxin of CpC. An exploratory analysis was performed to describe the prevalence and coexistence of the different antigenicity factors. From all the samples analyzed, 52 (64.2%) were positive for some of the antigenicity factors. From those 52 positive samples, the F4 fimbriae (44.4%) and LT toxin (40.7%) showed the highest positive prevalence (44.4%), and these two virulence factors coexisted in 24.7% of positive samples.

Implications: Fimbriae, specially the F4, and heat-labile toxin (LT) were highly detected on the samples analyzed and coexisted in 24.7% of all positive samples. Immunization against both antigens is imperative for optimal protection of piglets during their first days of life.

Post-weaning diarrhea in newly weaned pigs challenged with enterotoxigenic *E. coli*

Arndt, E.[1], Hanna, H.[1], Amezcua, M.[1], Friendship, R.[1] Farzan, A.[1,2]

[1]Department of Population Medicine, [2]Department of Pathobiology, University of Guelph, 50 Stone Rd E. Guelph, ON, N1G2W1
Email: afarzan@uoguelph.ca

Post-weaning *E. coli* diarrhea is an important cause of production losses and drug use on swine farms. Three trials were conducted to investigate the effects of infectious dosage, age, and pig susceptibility on post-weaning diarrhea associated with enterotoxigenic *E. coli* (ETEC). In each trial susceptible and resistant pigs were assigned to 4 groups including control (no challenge), low, moderate, and high dose (LD, MD, HD). Pigs were challenged at age of 15, 19, and 22 days in trial 1, 2, and 3, respectively. Pigs were monitored for diarrhea and rectal swabs were collected and tested for presence of ETEC bacteria quantitatively. All pigs were euthanized 5 days post-challenge. The presence of diarrhea was observed in 44.4, 55.6, and 61.1% of pigs in LD, MD, and HD group, respectively. There was no significant difference in diarrhea in susceptible (59.3%) and resistant (44.4%) pigs (p=0.28). However, a higher number of ETEC was detected in rectal swabs collected from susceptible pigs compared to resistant pigs (p=0.007). Regardless of infectious dose and susceptibility, 33.3, 44.4, and 77.8% of pigs had diarrhea associated with the challenge ETEC strain in trial 1, 2, and 3, respectively (p=0.02). Histopathology examination showed that 8.3% of resistant and 25% of susceptible pigs had intestinal adherent bacilli (p=0.12). Adherent bacilli were detected in 8.3, 25, and 33.3% of pigs in LD, MD, and HD group, respectively (p=0.11).

Implications: Diarrhea could be induced in pigs challenged with MD or HD at 3-weeks of age. However, some resistant pigs demonstrated diarrhea along with presence of intestinal adherent bacilli which might be associated with other microbial agents.

Acknowledgement: This work received financial support from Swine Innovation Porc, Ontario Pork, and Ontario Agri Food Innovation Alliance.

A challenge study to investigate the association between nursery diet complexity and *S. suis* disease

Hanna, E., Arndt, E., Amezcua, R., Friendship, R., Farzan, A.[1,2]

[1]Department of Population Medicine, [2]Department of Pathobiology, University of Guelph. 50 Stone Rd E. Guelph, ON, N1G2W1
Email: hannae@uoguelph.ca

In the nursery, economic losses due to reduced growth, and increased mortality and drug use occur as a result of *Streptococcus suis* disease. Nutrition has been well demonstrated to impact health of weaned pigs and may contribute to *S. suis* disease development. Low complexity diets for weaned pigs reduces feeding costs but may increase *Streptococcal* disease occurrence. This study examined the impact of diet complexity on *S. suis* disease in pigs experimentally infected with *S. suis*. In two trials, 3-week-old pig were weaned to a controlled facility (D1) and assigned to receive either standard nursery diet (high complexity; HC; n=30) or a less expensive low complexity (LC, n=30) diet with a higher inclusion of soybean meal in each phase and digestible protein ingredient restriction. Pigs were further assigned to one of three challenge groups: HC/control (HCC; no challenge), HC/S2 (challenged with serotype 2), HC/S9 (challenged with serotype 9), LC/control (LCC; no challenge), LC/S2 (challenged with serotype 2), and LC/S9 (challenged with serotype 9). Pigs were challenged with *S. suis* inoculum at D18, D24, and D30 (trial 1) and at D7 and D14 (trial 2) and assessed for clinical signs of *S. suis* disease and bacterial colonization for 6 wk (trial 1) and 4 wk (trial 2). Pigs in LC/S2 group had a lower occurrence of *S. suis* disease. Five (50%) and one (10%) of the pigs in HC/S2 and LC/S9 group developed clinical signs of disease. *S. suis* serotype 2 was detected from tonsil and/or nasal cavities of 5 (50%) pigs in HC/S2 group and 2 (20%) pigs in LC/S2 group ($p<0.05$). Significantly elevated post-challenge rectal temperature was observed in 3 of 4 challenged groups (Trial 1- HC/S2, LC/S2, and LC/S9 groups ($p<0.05$); Trial 2 - LC/S2 group ($p<0.05$).

Implications: Feeding low complexity nursery diet did not increase the risk of pigs becoming diseased with *S.suis* disease in this experimental challenge study suggesting that diet may not be a triggering factor for clinical *S. suis* disease.

Acknowledgement: This work received financial support from Swine Innovation Porc, Ontario Pork, Grand Valley Fortifiers, and Ontario Agri Food Innovation Alliance.

Early-life florfenicol administration has only temporary effects on the pig gut microbiome

Devin B. Holman[1] and Katherine E. Gzyl[1]

[1]Lacombe Research and Development Centre, Lacombe, AB T4L 1W1

This study investigated the effect of early-life treatment with florfenicol (Nuflor) (15 mg/kg) at 1 and 7 days of age on the short- and long-term development of the pig gut microbiome. Fecal samples were collected from all piglets in the control (n = 60) and treatment (n = 60) groups at 1, 4, 7, 11, 14, 21, 28, 84, 112, and 140 days of age. We also collected colostrum and milk samples from the sows at three different time points prior to weaning to associate this microbiome with the piglet gut microbiome. The microbiomes were characterized using 16S rRNA gene sequencing. A large shift in the composition of the gut microbiome was observed after florfenicol treatment on d 1 and 7 ($P < 0.05$). These changes were driven by an increase in the relative abundance of several members of the *Lachnospiraceae* family including *Blautia*, *Eisenbergiella*, *Epulopiscium*, and *Lachnoclostridium* spp., in the florfenicol-treated piglets. Additionally, there was a decrease in the relative abundance of *Actinomyces*, *Fusobacterium, and Peptostreptococcus* spp. compared to the untreated group. Microbial diversity was also significantly decreased after florfenicol administration. However, by d 21 just prior to weaning, the gut microbiome of both groups of piglets were largely similar in composition. The colostrum and milk collected from the sows was dominated by *Clostridium*, *Lactobacillus*, *Rothia*, *Staphylococcus*, and *Streptococcus*. Colostrum had a significantly higher relative abundance of *Acinetobacter*, *Escherichia*, and *Pseudomonas* spp. compared to milk collected 21 days post-farrowing. Milk from this collection time was enriched in *Actinobacillus*, *Bergeyella*, and *Gemella* spp. These results demonstrate the resilience of the piglet gut microbiome to early-life antibiotic exposure. In addition, we observed evidence of bacterial transfer from the sow to piglet via colostrum/milk.

Implications: This study shows that early use of florfenicol in piglets disturbs the bacteria in the piglet gastrointestinal tract only temporality. Piglets also did well short- and long-term (marketing weight) with or without antibiotic administration.

Immunological evaluation of an autogenous vaccine used in sows to protect piglets against *Streptococcus suis* infections

Alison Jeffery[1], Mélina Gilbert[1], Lorelei Corsaut[1], Simon Cloutier[2], Marie-Christine Frenette[2], Charles Surprenant[3], Marcelo Gottschalk[1] and Mariela Segura[1]

[1]Swine and Poultry Infectious Diseases Research Centre, Faculty of Veterinary Medicine, University of Montreal, St-Hyacinthe, Québec J2S 2M2.
[2]S.V.A. Triple-V Inc., Acton Vale, Québec J0H 1A0; [3]F. Ménard, Ange-Gardien, Québec.
E-mail: alison.jeffery@umontreal.ca

Streptococcus suis is an important bacterial pathogen in swine production. It causes major economic burden to the industry, mainly in the post-weaning period. *S. suis* is classified into 29 serotypes (i.e. antigenic variants) with diverse geographical distribution. There is no commercial vaccine available in North America. Therefore, autogenous vaccines are largely used as a preventive strategy; however, field studies on the immunological response (i.e. antibody production) induced by these vaccines are scarce. Previous studies performed in Canada showed that autogenous vaccines increase antibody levels against *S. suis* in vaccinated sows but the transfer of maternal immunity to piglets was either not improved or did not last longer than 18 days of age, with the vaccine formulations used in those studies. In this study, we assessed a sow vaccination program with an autogenous vaccine (manufactured by a company different from the previous studies) and containing five *S. suis* serotypes (2, 5, 7, 14 and 1/2). We evaluated the antibody response induced by the vaccine and transfer of maternal immunity from sows to their litters (until 7 week-old) on a commercial farm. Current results showed that vaccinated sows present higher levels of antibodies against the *S. suis* vaccine serotypes included in the vaccine than the non-vaccinated group. Maternal antibody transfer to their litters was higher in piglets born from vaccinated sows, which lasted until 3 to 5 weeks of age. Antibody functionality (i.e. capacity of antibodies to kill *S. suis*) was seen in one week-old piglets but then declined.

Implications: This study provides evidence that a sow vaccination program with herein studied autogenous vaccine increases maternal antibody levels in piglets up to 3 and/or 5 weeks of age depending on the serotype. More studies are required to fully characterize the clinical protective effect of this vaccine during the complete nursery period. This study also suggests that the autogenous vaccine manufacturing might affect the outcome of the vaccination program.

PRRS virus detection in dust collected with alternative aerosol collection tools

Christian Klopfenstein[1], Claudia Coulombe[1], Chantale Provost[3], Frédéric Fortin[1], Marie-Claude Poulin[2] and Carl A. Gagnon[3]

[1]Centre de Développement du Porc du Québec Inc. (CDPQ), Place de la Cité Tour Belle Cour, 450-2590 boulevard Laurier, Québec (Québec), G1V 4M6, Canada;_[2]Swine veterinarian consultant. [3]Centre de recherche en infectiologie porcine et avicole – FRQ, Faculté de médecine vétérinaire (FMV) de l'Université de Montréal. Email: cklopfenstein@cdpq.ca.

Aerosol transmission of Porcine Reproductive and Respiratory Syndrome virus (PRRSv) is a recognized transmission pathway for this virus. Finding the virus in aerosols is still a great challenge because collection efficiency with standard aerosol collection tools is known to be very low. Standard aerosol collection tools are defined in this paper as all those that are based on high-speed air flow aerosol impaction principle on either hard plates, filters, or liquids (see NIOSH Sampling and Characterization of Bioaerosols Manual for details of these methodologies). All the results discussed in this paper were obtained from aerosol collected close to pig pens (< 2 m) that were housing nursery piglets (n ≈ 15) that had been naturally exposed to PRRSv positive piglets for at least 7 days. Piglets were confirmed to be infected by PRRSv with a serological and oral fluid PCR test. CDPQ team explored the potential of two alternative custom aerosol collection tools: 1- low-cost aerosols sampler built on a computer cooling fan with low air speed (< 0.25 m/s) and high volume (> 4000 liters) with two different filters (XFAN + cotton gauze; XFAN + MERV11 membrane); 2- passive aerosol sampler based on one hour collection on a simple aluminum foil (ALU method). PRRSv was detected in the collected dust at the Molecular diagnostic laboratory at FMV. CDPQ team had high success rates with ALU (12/12 positive samples) and XFAN methodologies (17/24 positive samples). XFAN cotton gauze filter gave better results than XFAN MERV11 filter (10/12 versus 7/12). All samples (n = 36) were collected at different locations close (< 2m) to PRRSv infected pigs.

Implications: This research shows that it is possible to systematically find PRRS virus in aerosols collected within two meters of diseased animals using nonstandard aerosol collectors (XFAN and ALU). These low-cost highly efficient tools will be used to improve our knowledge of the aerosol PRRSv transmission.

Rapid, efficient and safe FTA card-based serotyping of *Actinobacillus pleuropneumoniae*

Oliver W. Stringer, Yanwen Li, Janine T. Bossé and Paul R. Langford

Section of Paediatric Infectious Diseases, Department of Infectious Diseases, Imperial College London, London, W2 1NY, United Kingdom
o.stringer18@imperial.ac.uk

Actinobacillus pleuropneumoniae (APP) is the causative agent of porcine pleuropneumonia, which results in significant economic losses worldwide due to decreased growth rates and high treatment costs. There are currently 19 known serovars of APP with certain serovars considered more pathogenic than others. Current vaccines against APP are bacterin based and do not give heterologous protection to all serovars. Therefore, seroprevalence is an important consideration when implementing effective vaccination strategies. Molecular serotyping of APP relies on the amplification of serovar specific capsule synthesis (*cps*) genes. Previously, we have described two multiplex polymerase chain reaction assays (mPCRs) for definitive serotyping of all 19 serovars of APP. However, mPCR serotyping requires potentially hazardous and infectious samples to be transported by specialist couriers to centralised laboratories, which can be expensive and discourages farms from engaging in routine testing programmes. Here we have developed a simple method to rapidly sample and process lung tissues utilising Whatman FTA cards. FTA cards are chemically coated to deactivate pathogens on contact and entrap nucleic acids, which are stable for at least 6-months at 37°C. With our optimised protocol we were able to produce a rapid (requires 2 x 5-minute washes) non-intensive (washed cards can be directly added to mPCRs) and chemical free (only requires water washes) method which results in no loss of sensitivity compared to the manufacturers protocol. Our FTA based isolation and possessing method displayed identical mPCR results to the gold-standard spin-column based tissue extraction methods, typically used in laboratories, in 85 clinical isolates and 22 lung samples.

Implications: We have shown that we can safely, rapidly and economically employ FTA cards for mPCR based serotyping of APP. We envision the uptake of FTA card sampling has the potential to equip farms with the tools required to rapidly respond to APP outbreaks and limit the economic losses incurred by this potentially catastrophic pathogen.

Why is there no vaccine for Swine Dysentery? B-cells: the missing link

Christine T. Yang[1], D.G. Ruwini Sulochana Kulathunga[2] and Matheus O. Costa[2]

[1]University of British Columbia, V6T 1Z4, [2]Western College of Veterinary Medicine, University of Saskatchewan, S7N 5B4
Email: ctyang@student.ubc.ca

Swine Dysentery (SD), first described in 1921, is characterized by mucohaemorrhagic diarrhea and colitis in pigs that severely impacts production and animal welfare. The causative agent is the spirochete *Brachyspira hyodysenteriae*. Despite its characterization 100 years ago, there are no effective forms of disease control other than antimicrobials, a point of concern as we enter the post-antibiotic era. Vaccines would be an attractive option to the industry, but so far all candidates failed to induce protection. A central point in vaccine development is the activation of B cells. Our objective here was to characterize the porcine B cell response to pathogenic *Brachyspira*. An immortalized porcine B-cell line was inoculated with pathogenic *B. hyodysenteriae* (n=2) cultured in broth at log phase and incubated together for 8 hours. Control B-cell cultures were sham inoculated (n=2). At the end of this period Brachyspira viability was verified using agar culture, and total RNA was extracted for cDNA synthesis. Expression of nine B-cell receptor activation pathways and synergistic cytokine genes were analyzed using Real-Time PCR. Our data revealed that infected B-cells exposed to the pathogen experienced a significant upregulation of the gene SLA-DRA (4.1 fold, *P*=0.003), when compared to controls. SLA-DRA helps the immune system recognize foreign proteins through peptide presentation. Interestingly, B-cell exposure to *Brachyspira* led to the downregulation of IFN-α (3.7 fold, *P*=0.0007) and IFN-β (4.3 fold, *P*=0.0003), when compared to controls. IFN-α/β increase B-cell survival and link innate and acquired immune responses. All other genes displayed no significant changes between infected and control groups. These results suggest that *Brachyspira* may supress B-cell activation, thus preventing the development of a humoral response.

Implications: Understanding the effects of *Brachyspira* on the host's immune system could contribute to future methods of vaccine development and SD treatment to prevent associated losses in the pork industry.

Implementation of prescription-platform vaccines for emerging and highly variable pathogens of swine

A. Young, Ph.D[1,2] and A. Petersen, M.S[2]

[1]Department of Veterinary and Biomedical Sciences, South Dakota State University, Brookings, SD. 57007. and [2]Medgene Labs, 1006 32nd Avenue, Brookings, SD. 57007

Vaccines are critical to support the Animal Agriculture industry. The current USDA regulatory process ensures a pure, safe, potent and efficacious product, but does take years and significant expense to develop and license. The USDA recently created new subcategories for platform and prescription (RxP) platform vaccines. These new regulatory pathways can now be used to rapidly respond with products that address pathogens with high diversity as well as new and emerging diseases, including rotavirus infections of commercial swine. Rotaviruses exist in multiple genotypes due to reassortment of viral genes. Current vaccines rely on outdated strains that may not cover current circulating strains. More flexible and modern vaccine technology is required to match the needs of producers. We have shown that a baculovirus-produced, rotavirus specific bivalent vaccine elicited high titers of neutralizing antibodies against target Rotavirus A. We have used bioinformatic approaches to define critical epitopes within the target proteins as evidenced by effective cross-neutralization in vitro, combined with in silico analysis of predicted protein structure. This work opened the door to the availability of targeted vaccines against circulating strains of Rotavirus A, and the potential for development of vaccines against Rotavirus B and C. Various field evaluation studies have been completed or are in-progress with swine integrators across the Midwest. Overall, clinical impressions from attending veterinarians at these swine facilities have been positive. Based on our work, we identified two proteins that together appear to provide high levels of neutralizing antibody. Further observation of the sequence differences between these virus proteins and the target construct identified several target regions that may correspond to target neutralizing epitopes.

Implications: These data indicate that the RxP approach can generate highly successful vaccines, and greater understanding of the nature of target epitopes within the viral proteins may extend the success of these vaccines against heterologous challenge.

A preliminary comparison of fecal microbiomes from early life to a week post weaning in commercial pigs across Canada

B. McCuaig[1], L. McCarthy[1], M. Arsenault[2], T. Prisnee[3], J. Harding[1], B. Lillie[2], V. Farzan[2], B. Willing[3], M. Links[1] and A. Van Kessel*[1]

[1]University of Saskatchewan, [2]University of Guelph, [3]University of Alberta
Email: andrew.vankessel@usask.ca

Although established as an important contributor to pig health and performance, high diversity and variability of the microbiome has challenged identification of a beneficial community composition that could be selected for and promoted. To address this challenge, we collected rectal swabs at intervals from birth to market for 4 pigs in each of 10 litters from 12 conventional commercial swine facilities and 7 facilities using raised without antibiotics (RWA) protocols from producers located in Quebec, Ontario, Manitoba, Saskatchewan, and Alberta. Here we report preliminary analysis of 16s rRNA gene sequence data (69,283,727 reads) from 1875 fecal samples collected at 4 ± 3 days of age, 1 day prior to weaning and 7 days after weaning. Raw sequences were processed using a DADA2 pipeline and resulting amplicon sequence variants assigned to 560 different genera. As expected, communities clustered uniquely by age at sampling, with greatest differentiation apparent in the post-weaning sample. At 4 days of age, fecal microbial profiles in AB and SK were distinguishable from samples collected from ON and QC. However, this separation was lost as the piglets aged such that province was not distinguishable 7 days post weaning. There was no apparent difference between RWA and conventional barns at any timepoint.

Implications: Further analysis will add collection time points and examine whether fecal community profiles can be associated with body weight, health score, and pig genotype.

The effect of limiting dietary non-essential amino acids on lysine requirement for protein deposition in growing pigs

Carley M. Camire (BScH)[1,2], Lucas A. Rodrigues (DVM, MSc)[1,2], Josiane C. Panisson (PhD)[1,2], Anna K. Shoveller (PhD)[3], Daniel A. Columbus (PhD)[1,2]

[1]Prairie Swine Centre, Inc., S7H 5N9, Saskatoon, Canada [2]Department of Animal and Poultry Science, University of Saskatchewan, S7N 5A8, Saskatoon, Canada [3]Department of Animal Biosciences, University of Guelph, N1G 2W1, Guelph, Ontario, Canada

Low protein (LP) diets supplemented with essential amino acids (EAA) fed to growing pigs reduces the dietary supply of excess amino acids (AA) and nitrogen (N) in the diet and reduces N excretion while maintaining growth performance. However, LP diets may not contain sufficient amounts of non-essential amino acids (NEAA) and N, potentially affecting utilization of EAA for protein deposition (PD). The objective of this study was to determine the effect of limiting dietary NEAA, measured as the EAA to total N ratio (EAA-N:TN), on the lysine (Lys) requirement for PD in growing pigs. A total of 80 growing barrows were randomly assigned to 1 of 10 dietary treatments (n = 8 pigs/treatment) in 8 blocks. Treatments were arranged in a 2 × 5 factorial design with factors of dietary EAA-N:TN (low vs. high) and 5 levels of dietary lysine (Lys) content (0.82, 0.92, 1.02, 1.12, and 1.22% standardized ileal digestible Lys). The low ratio (LR) diet had a ratio of 0.48, which is the ideal ratio based on previous studies, and the high ratio (HR) diet had a ratio of 0.55 which is considered limiting in NEAA or total N. Diets were fed at 2.8 × maintenance metabolizable energy requirements in 2 equal meals per day. Each experimental period consisted of a 7-d adaptation period and a 4-d collection period. Fecal and urine samples were collected each day, and nitrogen retention (NR) was determined as dietary N intake minus fecal and urinary N output. Protein deposition was determined as NR × 6.25. Data was analyzed as a MIXED model with fixed effects of ratio, Lys, and their interactions, and block as a random effect. Lys requirement was determined using two-phase linear or quadratic models. Protein deposition increased with increasing Lys levels regardless of the EAA-N:TN. Both PD and NR were maximized at 1.21% Lys level in HR diets, but no breakpoint was achieved for PD or NR in pigs fed LR diets. These results indicate that there is insufficient N in HR diets but not in LR diets, and that more EAA are required in LR diets to meet requirements for PD and improve NR.

Implications: These results suggest that NEAA need to be supplemented when the dietary EAA-N:TN is low and this formulation approach will improve the sustainability and profitability of pork production.

Soybean meal-based diets, effects on growth performance and health in nursery pigs

Amezcua, R., Arndt, E., Hanna, E., Friendship, R. and Farzan, A.

[1]Department of Population Medicine Department, University of Guelph. N1G 2W1
Email: mamezcua@uoguelph.ca

Expensive protein sources and antimicrobials are used to support piglet growth and gastrointestinal tract health during the first 2 week after weaning. Feed costs can be reduced by replacing these ingredients with plant protein such as soybean meal, but this ingredient may result in allergic reaction of the intestinal tract negatively impacting growth performance and health of the pigs. This study investigated the effect of low complexity (LC) diets on growth performance and health of pigs compared to pigs fed a high complexity (HC) standard diet. In four trials, weaned pigs were assigned to either one of two LC or a HC diets without in-feed antibiotics in the diets. The pigs receiving the HC diet in all four trials (n=180) were fed a four-phase diet (phase 1, 2 and 3 for one week each, and phase 4 for 2 weeks). In two trials, pigs were fed an LC diet (LC1, n=60) consisting of phase 3 (3 weeks,) and phase 4 (2 weeks) of the standard diet. In two trials, pigs were fed a customized LC diet (LC2, n=60) including phase 1,2 and 3 for one, 2 and 3 weeks, respectively. The LC diets contained a greater proportion of soybean meal than the corresponding HC diet at each phase. Growth, feed consumption, lameness, neurological signs, diarrhea, demeanor scores, and mortality, were recorded from weaning until 37 days after weaning. Rectal swabs were collected from a subset of pigs and tested for the presence of enterotoxigenic *E. coli* and S *suis.* Pigs on the LC diets grew slower (LC1: 170; LC2: 192 g/d) than pigs receiving the HC diet (222 g/d) ($P<.001$) during the first 14 days after weaning. Feed consumption and ADG from weaning to Day 37 was not different among diet groups (LC1: 464, and LC_2: 470 and HC: 483 g/d) ($P=.1$). Lameness mortality, *E. coli* and *S. suis* prevalence did not differ among diets. However, demeanour scores were poorer and diarrhea per pig per day was higher for the LC compared to the HC pigs ($P<.05$). Neurological signs of *S. suis* infection were observed in three pigs, all in the HC group.

Implications: Pigs fed LC diets had diarrhea and grew slower during the first 2 weeks after weaning. Compensatory growth was observed at Day 37. Mortality or the occurrence of health problems other than diarrhea did not differ among diets, which may encourage pig producers to utilize LC diets.

Acknowledgement: Swine Innovation Porc, Ontario Pork, GVF and OMAFRA

Effects of dietary supplementation with spray dried plasma, dried intestinal hydrolysates or bovine colostrum on the growth and intestinal health of weanling pigs

Luca Lo Verso, Philippe Tanguay, Marie-Pierre Létourneau Montminy and Frédéric Guay

Department of Animal Science, Laval University, Quebec, QC, G1V 0A6, Canada
Email: luca.lo-verso.1@ulaval.ca

In this study, we evaluated the potential of supplementing weaning diets with spray dried porcine plasma, dried porcine intestinal hydrolysates or bovine colostrum to modulate piglet growth, intestinal antioxidant status, immunity as well as remodelling and permeability. At weaning, 96 piglets were assigned to one of the following dietary treatments: 1) basal diet (CTRL); 2) basal diet containing spray dried porcine plasma at 50 g/kg feed (PP); 3) basal diet containing dried porcine intestinal hydrolysates at 50 g/kg feed (PIH); 4) basal diet containing bovine colostrum at 50 g/kg feed (BC). At d 0 and 14, growth performance was recorded; furthermore, eight pigs per treatment were euthanized at d 14 and intestinal samples were collected to evaluate the antioxidant status, immunity, cell permeability and remodelling. Our results showed that PP and BC supplementations increased the final weight and G:F ratio when compared to the other dietary treatments, whereas the ADG was higher in the BC group compared to PIH animals ($P<0.05$). Furthermore, feeding pigs with the PP diet affected the intestinal antioxidant status, as evidenced by the increased concentration of glutathione peroxidase when compared to the BC group ($P<0.05$). No dietary effect on the immune system or intestinal remodelling was recorded.

This study highlights the potential of dietary supplementations such as plasma proteins and bovine colostrum to act as a valid alternative to in-feed antibiotics, in order to reduce enteric diseases and maintain gut health and performance in in the post-weaning period.

Effects of feeding mid or zero tannin faba bean on energy, protein and amino acid, digestibility in growing pigs

P.W. Nyende, E. Beltranena, and R.T. Zijlstra

University of Alberta, Edmonton, Alberta, Canada T6G 2P5
Email: ruurd.zijlstra@ualberta.ca

Faba bean is an emerging, rotational pulse crop in western Canada grown to provide dietary starch and protein in swine diets. However, varying content of antinutritional factors (AFNs) such as tannins and vicine/covicine in faba bean cultivars may limit their inclusion in swine diets due to negative effects on energy and protein digestibility. To prioritize what cultivars to grow for incorporation in swine diets, four faba bean cultivars including two zero-tannin with low vicine/covicine (Snowbird and Snowdrop) and two mid-tannins (Fabelle with low vicine/covicine and Florent with high vicine/covicine) were evaluated for their effects on energy and amino acids digestibility. Eight 35-40 kg barrows fitted with ileal canula were fed four diets containing 95% of each of four faba bean cultivars for four periods. A nitrogen free diet containing 72% cornstarch was also fed to measure basal endogenous losses of protein and Amino acids. After 5 days of acclimation, faeces were collected for 2 days and ileal digesta for 2 days. On average, energy digestibility in pigs fed on diets with zero-tannin cultivars was 4.8% at the ileum and 5.7% for the total tract greater than pigs fed on diets with mid-tannin cultivars. However, digestible energy (DE) was lowest in pigs fed the diet with Florent, a mid-tannin cultivar (15.0MJ/kg) but did not differ from Snowdrop, a zero-tannin cultivar (15.4MJ/kg). Likewise, standardised ileal digestibility (SID) of total amino acid was significantly lower for mid-tannins (80%) than zero tannin faba bean cultivars (86%). Consequently, SID lysine (75.8%) and threonine (68.4%) were lowest in pigs fed on Florent cultivar while SID tryptophan for Florent diet (74.3%) did not differ from Snowbird (70.9%) and Snowdrop (71.9%), zero tannin cultivars. Diet with Fabelle, a mid tannin cultivar had the greatest SID for tryptophan (84.1%).

Implication: Mid-tannin cultivars had lower digestibility of both energy and amino acid than zero tannin cultivars, indicating a significant effect of tannins on nutrient digestibility. However, Florent cultivar which is high in both tannins and vicine/covicine had lower energy and amino acid digestibility than Fabelle which has high tannins but low in vicine/covicine content, indicating both AFNs may additively affect nutrient digestibility and thus should be considered for maximum incorporation of faba bean cultivars in swine diets.

Effect of nursery dietary protein source and functional amino acid supplementation on the response of pigs to a subsequent *Salmonella* challenge

Lucas A. Rodrigues[1,2], Josiane C. Panisson[1,2], Andrew G. Van Kessel[2,3] and Daniel A. Columbus[1,2]

[1]Prairie Swine Centre, Inc., Saskatoon, S7H 5N9, Canada, [2]Department of Animal and Poultry Science, University of Saskatchewan, Saskatoon, S7N 5A8, Canada, [3]The Vaccine and Infectious Disease Organization, University of Saskatchewan, Saskatoon, S7N 5E3, Canada
Email: dan.columbus@usask.ca

Feeding plant based (PB) nursery diets may aggravate the effects of disease in pigs, which may be attenuated by functional amino acid (FAA) supplementation. Thirty-two mixed-sex pigs (8.7 ± 0.23 kg) were assigned to 1 of 4 diet programs for 4 wk post-weaning. Treatments were arranged as a 2 × 2 factorial with dietary protein source (PB vs. animal-based [AB]) with basal (FAA–) or supplemented FAA profile (FAA+; Thr, Met, and Trp at 120% of requirements) as factors. Pigs were then placed onto a common grower diet and after a 7-d adaptation period, were inoculated with *Salmonella* Typhimurium (ST; 3.0×10^9 CFU/mL) and monitored for 7 d post-inoculation. Growth performance, fecal score, gut health, ST shedding score in feces, intestinal colonization and translocation were measured pre- and post-inoculation. Nursery fecal score was improved in AB- FAA+ pigs ($P < 0.05$) and post-inoculation fecal score was worse in PB compared to AB pigs ($P < 0.05$). Overall ST shedding, and cecal and colonic ST colonization were reduced in AB compared to PB pigs ($P < 0.05$). Translocation of ST to the spleen was decreased by FAA+ ($P < 0.05$). Post-inoculation, AB pigs had greater average daily gain compared to PB-FAA- ($P < 0.05$). Pigs fed AB-FAA- showed increased average daily feed intake compared to PB-FAA- pigs ($P < 0.05$). Gain:feed was increased in AB-FAA+ compared to PB-FAA- pigs ($P < 0.05$). Cecal myeloperoxidase was reduced in AB pigs compared to PB pigs ($P < 0.05$). Pigs fed plant-based nursery diets had increased susceptibility to an enteric challenge. Functional amino acid supplementation attenuated the negative effects of plant-based diets on the response of pigs to *Salmonella* challenge.

Implications: Feeding PB diets may result in increased pig susceptibility and a negative response to subsequent disease challenge. Supplementing PB diets with FAA above current requirements for growth may be a valuable nutritional strategy to improve the response of pigs to disease.

Effect of functional amino acid supplementation in nursery diets on the response of low and normal birth weight piglets to a subsequent *Salmonella* challenge

Lucas A. Rodrigues[1,2], Josiane C. Panisson[1,2], Livingstone A. Kpogo[2], Jolie C. González-Vega[4], John K. Htoo[4], Andrew G. Van Kessel[2,5], and Daniel A. Columbus[1,2]

[1]Prairie Swine Centre, Inc., Saskatoon, S7H 5N9, Canada, [2]Department of Animal and Poultry Science, University of Saskatchewan, Saskatoon, S7N 5A8, Canada, [4]Evonik Operations GmbH, Hanau-Wolfgang, 63457, Germany, [5]The Vaccine and Infectious Disease Organization, University of Saskatchewan, Saskatoon, S7N 5E3, Canada
Email: dan.columbus@usask.ca

Functional amino acid (FAA) supplementation attenuates the negative effects of disease challenge in pigs. The objective was to determine the effect of birth weight and FAA supplementation during a subsequent challenge. A total of 32 low (1.09±0.11 kg; LBW) and normal birth weight (1.58±0.11 kg; NBW) piglets were fed diets with a basal (FAA–) or supplemented FAA profile (FAA+; Thr, Met, and Trp at 120% of requirements) for 4 wk post-weaning. Pigs were then placed on a common grower diet and, after a 7-d adaptation period, were inoculated with *Salmonella* (ST; 2.2 × 10^9 CFU/mL) and monitored for 7 d post-inoculation. Growth performance, rectal temperature, gut health, ST shedding score in feces and translocation, and blood parameters were measured. Inoculation with ST increased rectal temperature, which was higher in LBW compared to NBW pigs ($P < 0.05$). *Salmonella* shedding and translocation to spleen were lower in NBW-FAA+ compared to NBW-FAA- pigs ($P < 0.05$). Post-inoculation average daily gain was higher in NBW-FAA+ compared to NBW-FAA- pigs ($P < 0.05$). The greatest haptoglobin was observed on d 4 post-inoculation in LBW-FAA- ($P < 0.05$). Serum antioxidant capacity increased in FAA+ pigs compared to FAA- ($P < 0.05$). Plasma superoxide dismutase was lower in LBW-FAA+ compared to LBW-FAA- pigs ($P < 0.05$). Reduced:oxidized glutathione at d 7 post-inoculation was higher in NBW compared to LBW pigs ($P < 0.05$). Cecal myeloperoxidase (MPO) was reduced in NBW-FAA+ compared to NBW-FAA- and colonic MPO was increased in LBW-FAA- ($P < 0.05$) compared to LBW-FAA+ pigs.

Implications: Low birth weight pigs may have increased susceptibility to subsequent disease challenge. Supplementing nursery diets with FAA above current requirements for growth is a nutritional strategy to support robustness of normal, but not low birth weight pigs.

Acid-binding capacity of feed in swine nutrition

L.F. Wang[1], J.R. Bergstrom[2], J.D. Hahn[2], M.G. Young[3] and R.T. Zijlstra[1]

[1]Department of Agricultural, Food and Nutritional Science, University of Alberta, Edmonton, Alberta, Canada T6G 2P5; [2]DSM Nutritional Products, Animal Nutrition and Health, Parsippany, NJ, U.S.A.; [3]Gowans Feed Consulting, Wainwright, AB, Canada T9W 1N3
Email: ruurd.zijlstra@ualberta.ca

Young pigs have an immature gut. In the stomach, weak acidity is a problem to digest nutrients properly and inhibit harmful bacteria; therefore, growth and health are reduced. Diets have an intrinsic ability to alter acidity in the stomach. Acid-binding capacity (ABC) can describe the ability of feedstuffs to increase or decrease acidity, thus influencing the pH of diets and the stomach. The ABC values vary among and within feedstuffs. Some minerals such as zinc oxide and limestone have high ABC, and thus worsen acidity in the stomach. Diets with high ABC might be detrimental to young pigs. Lowering diet ABC might be achieved by lowering protein content, using alternative minerals, grinding cereal grains coarsely, and including organic acids. Organic acids often have negative ABC values, and their dietary inclusion is effective to lower diet pH and ABC; however, their ability to increase gastric acidity has been inconsistent. Whether dietary organic acids inhibit hydrochloric acid secretion in the stomach requires clarification. Lowering diet ABC may increase digestion of protein and growth performance, but the efficacy of organic acids might be affected by other dietary factors such as mineral and fibre content, or ABC value of control diets. Dietary organic acids and their salts may function as alternatives to antibiotics and zinc oxide to maintain the gut health of pigs. Diets with high ABC were related to high incidence of post-weaning diarrhea. Dietary organic acids and their salts may increase microbiota diversity and reduce *E. coli* counts in the intestine, reduce diarrhea incidence and *Salmonella* prevalence, and reduce ammonia emissions. However, dietary organic acids may affect feed preference and palatability, cause metabolic acidosis, and affect bone health of pigs following long term use.

Implications: Feeding diets with low ABC may increase nutrient digestion and gut health of pigs. Minerals with high ABC should be minimized in diets for young pigs. Organic acids are effective in lowering ABC of diets fed to young pigs.

Maternally derived antibodies test – MDA TT: tool to evaluate the colostrum intake

Mike DeGroot[1], Joaquin Miguel[2], Irene Galé[2]

[1]HIPRA Canada, (K1Y4S1) Ottawa (Canada), [2]HIPRA HQ, 17170 Amer, (Spain).
Email: mike.degroot@hipra.com

Colostrum intake is a crucial point in the health management of swine farms. Previous studies have directly correlated it to different productive parameters in piglets, such as pre-weaning mortality. Colostrum uptake can be measured in piglets by different methods: qualitative as the thermography, the difference in piglet weight between farrowing and 24 hours later or, qualitative as the inmunocrit. All these techniques vary on labour, cost investment and, especially sensitivity and robustness. Considering that the vaccination rate against Swine Erysipelas (SE) is close to 100% in sows and SE titers are transferred through colostrum (maternally derived immunity (MDI)). These titers can be quantified in both sows and piglets using a SE indirect ELISA kit, with robust results. Therefore, a direct correlation between a sow's titer and her offspring's titers is expected in the case of good colostrum intake. The aim of this publication is to introduce in Canada a service to evaluate colostrum intake. An algorithm (MDA TT) offers a score of the quality of immune transfer, objectively and precisely, considering the previously commented correlation between the sow and piglet SE's titers, measured at day 7 after farrowing. This algorithm is based on the study carried out in more than 150 farms located in different countries. The correlation between the SE antibody titers of the sows and the titers from their respective piglets was R=0.66 (*p-value < 0.001*), it was also found at each farm level. The algorithm used elicits a score that provides an objective value for the colostrum intake. The score is in a range from 0 to 100%. Hence, the MDA TT allows you to identify failures in the management of the colostrum intake and improve it. For example, in piglets, certain diseases such as neonatal diarrhea or atrophic rhinitis are controlled through the transfer of maternal antibodies. This tool could help to discern if the incidence of these diseases is related to poor colostrum intake (low MDA TT score) and consequently a lack of immunity in piglets, or a lack of efficacy of the vaccines to elicit immunity (high MDA TT score).

Implications: The MDA TT is an easy, fast, and objective way to evaluate the piglet colostrum intake in farms, giving the possibility to identify failures in this process, and consequently work on its improvement.

Evaluating piglet welfare after long and short transport using behavioural measures

H.R. Golightly,[1] J. Brown,[2] R. Bergeron,[3] Z. Poljak,[1] and T.L. O'Sullivan[1]

[1]Department of Population Medicine, Ontario Veterinary College, University of Guelph, ON, N1G 2W1. [2]Prairie Swine Centre, Saskatoon, SK, S7H 5N9. [3]Department of Animal Biosciences, University of Guelph.
Email: hgolight@uoguelph.ca, tosulliv@uoguelph.ca, jab651@mail.usask.ca

Behavioural measures are important for assessing animal welfare as they complement physiological changes in the evaluation of states of discomfort, hunger, and thirst. The objective of this study was to measure the effects of transportation on piglet welfare by comparing the behaviour of piglets after long (LD: >30h) or short (SD: <3h) transport events. Piglets were 17-24 days old at the time of transport; LD piglets were weaned ≤ 6 days before transport and SD piglets were weaned the morning of transport. The LD transport events used a potbelly trailer and originated from a sow barn in Saskatchewan, while the SD transport events used a flat deck trailer and originated from a sow barn in Ontario. All transports delivered piglets to multiple nursery barns in Ontario. Video observations of piglets in nursery pens were collected for 5.5h immediately after arrival (T1), and again three to four days after arrival (T2), with 240 LD piglets and 200 SD piglets observed, respectively. Lying, sitting, drinking, and feeding behaviours were recorded and mixed linear regression models were used to evaluate the effect of transport duration on piglet behaviour after transport. When considering all piglets, time spent (% of observations) lying was greater than any other observed behaviour on both sample days (T1: 77.3%, T2: 74.2%). The time spent feeding (T1: 11.1%, T2: 13.8%) and drinking (T1: 8.7%, T2: 7.6%) were much lower in comparison, and sitting behaviour (T1: 8.7%, T2: 7.1%) was observed the least often. When comparing groups by transport duration, LD and SD piglets spent similar amounts of time lying down at T1 but SD piglets spent more time lying down at T2 than LD piglets (86.8% vs 61.6%, $P<0.01$). LD piglets spent more time feeding (T1: 24.3% vs 5.1% $P<0.01$, T2: 19.8% vs 9.6% $P<0.01$) and drinking (T1: 13.6% vs 5.6% $P<0.01$, T2: 10.2% vs 5.7% $P<0.01$) than SD piglets at both timepoints. Similarly, LD piglets spent more time sitting than SD piglets at both T1 (13.8% vs 5.4% $P<0.01$) and T2 (9.5% vs 5.3% $P<0.05$).

Implications: The time spent lying, sitting, feeding, and drinking were different between the LD and SD piglets observed after transport. Increased feeding and drinking were displayed by LD piglets and may be explained by their earlier weaning before transport. However, LD piglets may also benefit from mitigations that reduce time without feed and water during transit therefore reducing hunger and thirst.

Examining sow anemia and the incidence of stillbirths

Sarah V. Hill[1], Maria Amezcua[1], Eduardo Ribeiro[2], Terri O'Sullivan[1] and Robert M. Friendship[1]

[1]Department of Population Medicine, [2]Department of Animal Bioscience
University of Guelph, 50 Stone Rd E, Guelph, ON N1G 2W1

Based on 2020 Canadian pork benchmark data, the average stillborn pigs/litter is 1.26 (PigCHAMP, 2021). The rate of stillbirths has been increasing at a steady rate over the last decade. Despite the extensive research performed examining this issue, the rate of stillbirths has increased over the past decade. As part of a larger randomized control trial examining risk factors associated with stillbirths, a subset of 96 sows have been further assessed for low hemoglobin (Hb) levels in order to assess whether sow anemia is associated with an increased risk of stillbirths. The median parity of sows in this study was 2.5, ranging from gilts to parity 7 sows. To analyze the Hb level in sows prior to farrowing, a blood sample was collected into a tube (BD Vacutainer, BD, Franklin Lakes, NJ, USA) containing an anticoagulant (EDTA) and stored in an insulated container before being submitted to the Animal Health Laboratory (AHL), University of Guelph within 3 hours after collection. Samples were analyzed for hemoglobin determination and a complete blood count using an ADVIA 2120/2120i Hematology System (Siemens Healthcare Diagnostics, Deerfield, IL, USA). Current data indicates the average piglets born alive was 12 (+/- 3) with average stillbirths/litter was 1.1 (+/- 1.63). A total of 48 sows (50% of sows) had at least 1 stillbirth (min 0, max 7). The average parity for sows with stillbirths was 2.85, while the average parity of those with no stillbirths was 2.19. Sows were considered borderline anemic if Hb concentrations were between 90 and110 g/L. A total of 14 sows (14.5%) were considered borderline anemic. Further data collection and analysis is ongoing.

Implications: There are both welfare and economic implications associated with high stillbirth rates; the ability to understand and decrease the stillbirth rate through targeted management factors such as addressing anemia is important.

Acknowledgements: This research has been funded by Ontario AgriFood Innovation Alliance and Grand Valley Fortifiers.

Does feeding time affect farrowing time in sows?

Josiane C. Panisson (PhD)[1,2], Hyatt Frobose (PhD)[3], Jennifer Brown (PhD)[1,2] and Daniel A. Columbus (PhD)[1,2]

[1]Prairie Swine Centre, Inc., S7H 5N9, Saskatoon, Canada
[2]Department of Animal and Poultry Science, University of Saskatchewan, S7N 5A8, Saskatoon, Canada.
[3]Jyga Technologies, Inc., 66032, Garnett, KS, USA
Email: dan.columbus@usask.ca.

It is sometimes necessary to provide assistance to sows and piglets during farrowing in order to improve piglet survival, resulting in increased labour costs. There is evidence in cattle and ewes that feeding at night results in an increase in the incidence of calvings during the day. In commercial farms, sows are typically fed during daytime hours, with a high percentage of farrowings occurring unassisted overnight. The objective of this study was to determine the effect of feeding time (morning vs evening) on farrowing time in sows. A total of 278 sows were used in groups of 12 sows/block and 11 blocks/treatment. Sows were assigned to either morning (0700 h) or evening (1900 h) initiation of feeding upon entry to farrowing room (~d110 of gestation). All sows were placed on a common lactation diet and fed according to established sow feeding curves with feed intake controlled and monitored automatically (Gestal Solo, JygaTechnologies). All sows were switched to a 0700 h feeding time post-farrow. Cameras were placed in each create to allow for recording of all farrowings. Initiation of farrowing and farrowing duration were collected. Total number of piglets born (alive (A) and stillborn (S)) and pre-weaning mortality (PWM) was recorded. There was no difference in A, S, or PWM ($P > 0.10$). Initiating sow feeding in the morning increased the frequency of farrowing during the day (7:00 am to 3:00 pm) compared to evening fed sows ($P < 0.05$; 56.19% *vs* 30.39%, respectively). Morning feeding also tended to result in a shorter farrowing duration ($P < 0.10$; 4:14 vs 4:54 h).

Implications: These results suggest that the initiating feeding in the morning may be beneficial for maximizing the number of farrowings that occur and are completed when workers are present.

Effects of heat lamp location and piglet placement at birth on body temperature, suckling behaviors and survival

Aridany Suarez-Trujillo[1], Alicia Denton[2], Alton Holstine[2], Kara Stewart[2]

[1]Berry College, 2277 Martha Berry Highway NW, Mount Berry, GA 30149 [2]Purdue University, 270 South Russell St., West Lafayette, IN 47907
Email: krstewart@purdue.edu

Heat lamps are extensively used as a heat source for newborn piglets in the swine industry. The assistance to piglets to reach the udder is variable between operations. The objective of this study was to evaluate the effect of the location of the heat lamp (HL) and piglets (PL) at birth on the piglet's suckling behavior, body temperature, growth, and survival. A total of 1053 piglets from 87 litters were used for this study. Litters were blocked by parity and randomly assigned to one of the four treatments in a 2x2 factorial design, combining the HL and PL at birth. The four treatments were: HLU/PU, HL situated in the sow's udder, and piglets placed at the sow's udder; HLB/PU, HL at the sow's back and piglets at the udder; HLU/PV, HL at the udder and piglets placed at the sow's vulva; and HLB/PV, HL at the sow's back and piglets at the vulva. After birth, each piglet was weighed and rectal temperature taken. Body temperature was recorded at 30 min, 1h, and 2 h. The time from birth to first approach to the udder and first suckle were recorded. Body weight was recorded at 24h, and weekly until weaning. Piglet mortality and reasoning for death was recorded through weaning. Data was analyzed as a complete factorial with 4 treatments in addition to the effects of HL and PL evaluated separately using PROC MIXED and GLIMMIX in SAS. HLB/PV piglets tended to be colder at 30 min after birth ($P=0.097$) than the other treatments. When the individual effect of PL was analyzed, piglets placed at the sow's vulva had lower temperatures at 30 min ($P=0.019$) than piglets at the udder. The time taken to reach the udder and for the first suckle were greater ($P < 0.001$ and $P=0.019$, respectively) in piglets placed at the vulva. Mortality, colostrum intake, body temperature after 30min, and growth through weaning were not affected by treatment.

Implications: Placing the piglets close to the udder after birth can minimize decrease in piglet temperature after birth and decrease the time from birth to suckling. However, additional measures are needed to improve colostrum intake in newborn piglets.

Influence of carbohydrate rich co-products and fiber on growth performance and fat quality in finishing pigs

Ferial Amira Slim, Luca Lo Verso and Frédéric Guay

Laval University, Department of Animal Science, Québec, QC, G1V 0A
ferial-amira.slim.1@ulaval.ca

In recent years, different strategies have been undertaken to reduce feed cost including incorporation of co-product. However, addition of large quantities of co-product can influence meat quality, especially fat quality.

The main objective of the present study was to assess the effect of feeding high dairy/carbohydrate diets on growth performance and fat quality in finishing pigs. A total of 36 pigs ((Yorkshire x Landrace (x Duroc)) were raised in commercial diets up to 50 kg. Pigs were distributed in six dietary treatments for 4 weeks. The first three treatments were: a control diet (C) consisting of a corn-soybean meal diet, 25CoP containing 25% co-products (sucrose, whey and bakery meal) and 50CoP containing 50% co-product. Three other diets were formulated from 50CoP and supplemented with three sources of fiber: oat hull (50Oat), soy hull (50Soy) and wheat middling (50 Midd). Feed intake and average daily gain were recorded. After 4 weeks, pigs were slaughtered and a 5-cm strip of backfat was taken from each animal. A measurement of thermodynamic characteristics (melting behavior and solid fat content) were determined by differential scanning calorimetry. ADG and ADFI were reduced in 25CoP compared to 50CoP with an intermediary value for C ($P<0.05$). 50Soy had lower ADFI while 50Oat had higher feed conversion than 50CoP ($P<0.05$). Only meat lean yield tended to be increased in 25CoP and 50Soy compared to 50CoP ($P<0.09$). Melting point onset and maximum were higher in 50CoP compared to C and 25CoP ($P<0.05$). Fiber supplementation reduced these melting points compared to 50CoP ($P<0.01$). Proportion of fat melted at -5°C and 5°C was decreased in 50CoP compared with C and 25CoP. Fiber supplement increased these proportions of melted fat ($P<0.01$).

Implications: This project shows that diet containing around 50% of rich carbohydrate co-product can negatively affect fat quality and fiber supplementation to reach 12% NDF restore the fat quality.

Genome-wide association analyses (GWA) for carcass and meat quality traits in pigs using imputed whole-genome sequence (WGS)

Marzieh Heidaritabar[1], Abe Huisman[2], Marco Bink[2], Patrick Charagu[3] and Graham Plastow[1]

[1] Department of Agricultural, Food and Nutritional Science, University of Alberta, Livestock Gentec, College Plaza, 8215 112 Street, Edmonton AB, T6G 2C8, [2] Hendrix Genetics Research, Technology & Services B.V., Boxmeer, the Netherlands, [3] Hendrix Genetics, Swine Business Unit, Regina, SK, Canada

Pork quality is an economically important trait in the swine industry, and it is one of the main breeding objectives in today's pig breeding programs. Understanding the complex genetic mechanisms underlying carcass and meat quality traits is a priority for genetic improvement. To date, most GWA studies have used medium to high-density marker panels to detect the genomic regions associated with carcass and meat quality traits. Use of WGS is expected to improve identification of associated regions. We performed GWA analyses for several meat and carcass quality traits, including 18 meat color traits (Minolta L*, a*, and b* for fat, quadriceps femoris muscle, thawed loin muscle, ham gluteus medius, ham gluteus medius, and longissimus dorsi muscle on the fresh loin), peak shear force, fresh muscle pH, drip loss, marbling score, and several belly traits. Our analyses were conducted for two datasets: 954 crossbreds (CBs) and 891 purebreds (PBs). Sequence markers, called across the 60 sequenced animals, were imputed from a medium-density marker panel (~ 44K) in all genotyped animals. We applied single marker association analysis accounted for polygenic effects through the genomic relationship matrix for each dataset. Our objective was to investigate whether the use of WGS will detect more associated regions in pigs compared with lower density marker panels. Our findings showed that more associated regions were detected for traits in PBs than those of CBs. Generally, compared to lower density panels, using WGS did not lead to the identification of novel genomic regions, except for a few traits in PBs. Novel detected associated regions included 3 genomic regions for belly bend on SSC7 (at 111.48-118.67Mbp), SSC8 (at 124.05-124.13Mbp), and SSC13 (at 110.97-123.61Mbp), 2 regions for fresh muscle pH on SSC2 (at 146.65-147.87Mbp), and SSC6 (at 168.26-168.33Mbp), one region on SSC2 (at 142.79-144.77Mbp) for longissimus dorsi muscle Minolta b*, and one region on SSC10 (at 38.12-38.28Mbp) for longissimus dorsi muscle Minolta L*. These results provide additional insights for the genomic regions affecting meat and quality traits.

Implications: The use of WGS for GWA studies is presently of limited value, likely due to power issues. The added value of WGS relative to lower density marker panels when performing GWA studies should be explored in larger pig populations, as the low number of sequenced pigs (60) and low number of genotyped and phenotyped individuals (891-954) may have limited the potential advantage of WGS to increase the power of GWA analyses.

Pork meat and belly qualities as influenced by feed efficiency selection in a sire and a dam line

A. Saikia[1], G. Mejicanos[1], J. Rothy[2], E. Rajendiran[1], C. Yang[1], M, Nyachoti[1], H. Lei[3], E. knol[4], R. Bergsma[4], Y. Wu[1], S. Jin[1] and A. Rodas-Gonzalez[1]

[1]Department of Animal Science and [2]Food Human Nutritional Sciences, University of Manitoba, Winnipeg, MB, Canada, R3T 2N2; [3]Topigs Norsvin Canada Inc., Oak Bluff, MB R4G 0C4; [4]Topigs Norsvin Research Centre, Beuningen, the Netherlands

Concern has been expressed on pigs selected for high feed efficiency because they may produce pork with inferior quality; however, contradictory results have been shown to date. One hundred boars slaughtered at approximately 115 kg of body weight were used to study the influence of genetic lines (dam-line vs sire-line) and estimated breeding value for feed conversion (EBV-FC; low-, intermediate- and high-efficient groups) within the Large White breed on carcass attributes, meat and belly qualities. At 96 h *postmortem*, left carcass sides were evaluated for backfat depth and thickness (on first rib, last rib and last lumbar), loin depth and area, subjective muscle colour and marbling, then fabricated into primals, and finally dissected into fat, bone, and lean to evaluate the yield as part of the live Computed Tomography scan validation project. Boneless loins and skin-on boneless bellies were obtained from the right carcasses. Loin pork chops (2.5-cm thick) were used for Warner Bratzler shear force (WBSF), cooking traits, drip loss, pH and objective colour evaluation. Belly dimension (length, width and thick) and belly firmness (subjective belly firmness score and belly-flop angle) were measured. Slaughter weight was included as a covariate in the model due to the slaughter weight difference among genetic lines. No difference in carcass weight was detected among genetic lines ($P > 0.05$); however, the sire-line had a greater loin area and loin depth and thinner fat depth than the dam-line ($P < 0.05$), which favoured higher lean and lower trimmed fat proportions in the sire-line ($P < 0.01$). For quality, genetic lines expressed minimal colour changes and drip losses ($P < 0.05$), with no differences in pH, marbling level and cooking traits ($P > 0.05$). On the other hand, regardless of genetic line, high-efficient animals presented the greatest loin area (> 1.16 cm^2), the deepest loin (> 2 mm), the thinnest back fat on the last rib and lumbar level (< 4 mm) ($P < 0.05$), the highest lean yield (> 3%) and the lowest proportion of trimmed fat (< 2.5%) ($P < 0.01$) compared with other efficient groups. For meat quality, efficient groups did not differ on pH, marbling level, drip loss, objective and subjective colour score ($P > 0.05$). However, the interaction genetic line x EBV-FC affected WBSF, where sire-line and intermediate-efficient groups had the lowest WBSF (3.5 kg) ($P < 0.05$). Belly dimension and firmness did not differ among genetic lines and EBV-FC groups ($P > 0.05$).

Implications: Based on the advantageous performance observed in most carcass yield traits, high-efficient animals offer a favourable response in greater loin and leaner animals, without decreased meat and belly quality traits.

Banff Pork Seminar
Dept AFNS, University of Alberta
4-10 Agriculture Forestry Centre
Edmonton, AB T6G 2P5